National Curriculum

Mathematics
Attainment Tests

Key Stage 3
(11–14 year olds)

STAFFORD BURNDRED
B.A. (Hons), M.Ed., Adv.Dip.Ed.

Mathematics National Curriculum Co-ordinator
at Henry Harbin School, Poole

DP PUBLICATIONS LTD
Aldine House, Aldine Place,
142/144 Uxbridge Road,
Shepherds Bush Green,
London W12 8AW
1992

For my wife, Carolyn.

A catalogue record for this book is available from the British Library.

ISBN 1 870941 63 2

Copyright DP Publications © 1992
Text copyright Stafford Burndred © 1992

First Edition 1992
Reprinted 1992

All rights reserved
No part of this publication may be reproduced, stored in a retrieval system, or transmitted in any form or by any means, electronic, mechanical, photocopying, recording, or otherwise, without the prior permission of the copyright owner.

Typeset by DP Publications Ltd

Printed by
The Guernsey Press Co Ltd
Braye Road, Vale
Guernsey, Channel Islands

Contents

Preface		v
Introduction		vi
How to use this book		vii
Number		1
Level 4	Attainment Test 1	10
	Attainment Test 2	11
	Attainment Test 3	12
Level 5	Attainment Test 1	20
	Attainment Test 2	21
	Attainment Test 3	22
Level 6	Attainment Test 1	28
	Attainment Test 2	29
Level 7	Attainment Test 1	35
	Attainment Test 2	36
Algebra		37
Level 4	Attainment Test 4	41
Level 5	Attainment Test 4	47
	Attainment Test 5	48
Level 6	Attainment Test 3	52
Level 7	Attainment Test 3	60
	Attainment Test 4	61
Shape and Space		63
Level 4	Attainment Test 5	68
Level 5	Attainment Test 6	75
	Attainment Test 7	77
Level 6	Attainment Test 4	83
Level 7	Attainment Test 5	88
Handling Data		89
Level 4	Attainment Test 6	97
Level 5	Attainment Test 8	105
Level 6	Attainment Test 5	112
Level 7	Attainment Test 6	121

Contents

Using and Applying Mathematics 123
    Section A   Using Mathematics   124
    Section B   Mathematical Communication   125
    Section C   Mathematical Reasoning   125
    Section D   An Example of Problem Solving   127
    Section E   Practical Tasks   128
                   Suggested solutions   130
    Section F   Mathematical Reasoning   133
                   Suggested solutions   135

Appendices 137
    Appendix 1   A mental arithmetic test for each level   138
                      Solutions to mental arithmetic tests   140
    Appendix 2   Answers and mark scheme for each test   141
                      (including pass mark)
    Appendix 3   Three one-hour mock examinations for
                      each of Levels 5 and 6.   156
    Appendix 4   Answers to mock examinations   173

Plot your progress   Inside back cover

Preface

Aim
This book enables anyone between the ages of 11 and 14 to know what level he or she has reached in mathematics, in relation to the standards laid down within the National Curriculum; and to help him or her progress through **Key Stage 3**.

Need
The **National Curriculum** – Average pupils aged 11 are expected to be capable of achieving Level 4 in the specified attainment targets. In their subsequent three years, they are expected to have moved on to achieve Level 6. High fliers may be able to achieve Level 8.

It is therefore useful for all 11–14 year olds to know where they are at the moment, in order to appreciate what is expected of them in the National Curriculum, and have a firm foundation for their GCSE years.

Approach
The book is grouped by the topics for which Attainment Targets and Programmes of Study have been laid down – ie Number, Algebra, Shape and Space, Handling Data and Using and Applying Maths.

Within each topic, the **achievements** and **skills/knowledge** required are given, and half hour **attainment tests** applicable to each of the Levels 4, 5, 6 and 7 are supplied.

Note: In the main, there is one half hour test for each level. Where the content is too great to be covered in one half hour test, a second or even third test is provided.

Answers, with a **marking scheme**, are given in Appendix 2.

Appendix 1 contains a mental arithmetic test for each level, together with solutions. In Appendix 3, there are 'mock examinations' for Levels 5 and 6, which cover all the topics. The answers to these can be found in Appendix 4.

Introduction

A Brief Explanation of the National Curriculum for Mathematics

The National Curriculum for Mathematics is divided into five main topics: Number; Algebra; Shape and Space; Handling Data; and Using and Applying Maths. Each topic has ten levels of attainment.

Level 1 is the work which a pupil beginning school at 5 years of age would be able to attempt. Level 10 is the level which the most able pupils aged 16 would be able to attempt. The range of levels for pupils aged 11–14 (ie Key Stage 3) is given as Levels 3 to 8. However, the vast majority of pupils will be covered by Levels 4 to 7, and this book concentrates on these levels.

- The testing of pupils during Key Stage 3 will take two main forms, firstly the school-based testing which will be administered by schools, generally during lessons, and secondly the official tests which will be distributed by the DES.
- This book contains guidance for pupils and parents on the requirements of the National Curriculum, together with tests at each level for each topic. Practical sample tests, coursework and mental arithmetic tests are also included.

This table shows the ages at which a pupil should complete the given Attainment Levels of Key Stage 3.

National Curriculum Attainment Levels	Pupil should complete this level at the age of	Pupil should complete this level by the end of school year*
4	11	6 (age 10–11)
5	13	8 (age 12–13)
6	15	10 (age 14–15)

* In future pupils will not be referred to as 'first year secondary' and so on. Instead the following will be used: Year 1 age 5–6; Year 2 age 6–7; Year 3 age 7–8; etc.

An average pupil should take two years to complete each of these attainment levels. Therefore a pupil will reach the end of Key Stage 3 at the age of 14 (year 9 age 13–14).

- It can be seen from the table that an average pupil aged 14 should have completed Level 5 and about half of Level 6.

How to use this book

You should first attempt the Level 4 tests from each of the topic chapters – **Number, Algebra, Shape and Space, Handling Data** – and then attempt the tasks in the topic **Using and Applying Mathematics**. Each test has a mark scheme and a pass mark (Appendix 2). If you pass a test, you should record this on the Progress Chart (inside back cover of this book).

When you have passed every test for Level 4, including the mental arithmetic test in Appendix 1, then you have successfully achieved Level 4. If you fail any test, this indicates an area of weakness where you will require further study. When you have completed a level, you should progress to the tests at the next level.

- Appendix 2 contains answers, a mark scheme and a pass mark for each test. The tests in this book, as is the case with the National Curriculum tests, are criterion-referenced. This means that you need only achieve a pass mark to pass. If the pass mark were 70% then a pupil who obtained 70% would gain a pass and a pupil who obtained 100% would also gain the same pass. (Under the old system of norm-referencing the pupil obtaining 100% would have obtained a higher grade.)

- Appendix 3 contains mock examinations for Levels 5 and 6. Each examination lasts for one hour and will provide valuable practice in examination technique.

Overleaf is a annotated picture of a typical page which explains what each symbol means, and how you should work through each test question.

How to use this book

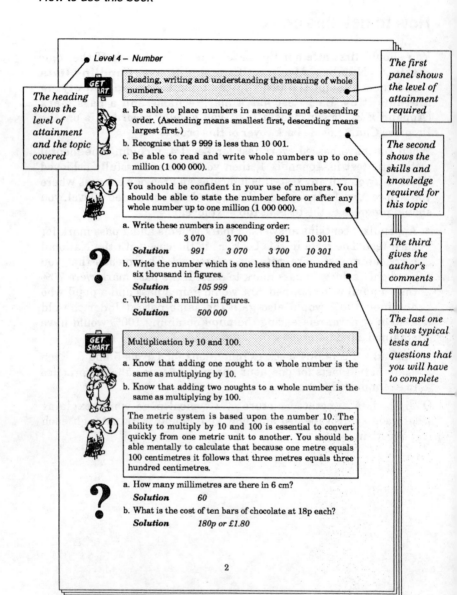

Level 4 – Number

Reading, writing and understanding the meaning of whole numbers.

a. Be able to place numbers in ascending and descending order. (Ascending means smallest first, descending means largest first.)
b. Recognise that 9 999 is less than 10 001.
c. Be able to read and write whole numbers up to one million (1 000 000).

You should be confident in your use of numbers. You should be able to state the number before or after any whole number up to one million (1 000 000).

a. Write these numbers in ascending order:

        3 070   3 700   991   10 301

   Solution   *991*   *3 070*   *3 700*   *10 301*

b. Write the number which is one less than one hundred and six thousand in figures.

   Solution   *105 999*

c. Write half a million in figures.

   Solution   *500 000*

Multiplication by 10 and 100.

a. Know that adding one nought to a whole number is the same as multiplying by 10.
b. Know that adding two noughts to a whole number is the same as multiplying by 100.

The metric system is based upon the number 10. The ability to multiply by 10 and 100 is essential to convert quickly from one metric unit to another. You should be able mentally to calculate that because one metre equals 100 centimetres it follows that three metres equals three hundred centimetres.

a. How many millimetres are there in 6 cm?

   Solution   60

b. What is the cost of ten bars of chocolate at 18p each?

   Solution   *180p or £1.80*

The heading shows the level of attainment and the topic covered

The first panel shows the level of attainment required

The second shows the skills and knowledge required for this topic

The third gives the author's comments

The last one shows typical tests and questions that you will have to complete

On the first page of every chapter you will find a symbol key to remind you what each box contains.

Number

The ability to use and understand number is as important as the ability to read and write. We need number to put objects into an order. How would you find a house number 372 unless you knew that it was higher than 370 but lower than 374? There are many other everyday examples of the use of number – time and money are two examples which you probably use several times each day.

Key

 The blackboard represents the level of attainment required.

 The dog holding a calculator shows the skills and knowledge required for this topic.

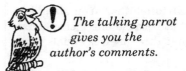 The talking parrot gives you the author's comments.

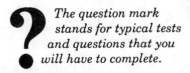 The question mark stands for typical tests and questions that you will have to complete.

Level 4 – Number

Reading, writing and understanding the meaning of whole numbers.

a. Be able to place numbers in ascending and descending order. (Ascending means smallest first, descending means largest first.)
b. Recognise that 9 999 is less than 10 001.
c. Be able to read and write whole numbers up to one million (1 000 000).

You should be confident in your use of numbers. You should be able to state the number before or after any whole number up to one million (1 000 000).

a. Write these numbers in ascending order:

              3 070   3 700   991   10 301
Solution   991   3 070   3 700   10 301

b. Write the number which is one less than one hundred and six thousand in figures.
Solution   105 999
c. Write half a million in figures.
Solution   500 000

Multiplication by 10 and 100.

a. Know that adding one nought to a whole number is the same as multiplying by 10.
b. Know that adding two noughts to a whole number is the same as multiplying by 100.

The metric system is based upon the number 10. The ability to multiply by 10 and 100 is essential to convert quickly from one metric unit to another. You should be able mentally to calculate that because one metre equals 100 centimetres it follows that three metres equals three hundred centimetres.

a. How many millimetres are there in 6 cm?
Solution   60
b. What is the cost of ten bars of chocolate at 18p each?
Solution   180p or £1.80

Level 4 – Number

> Use and understand decimal numbers. Read scales with decimal numbers.

a. Know that the number in the first decimal place is tenths, the number in the second decimal place is hundredths.
b. Be able to read scales, eg
   be able to read lengths from a ruler
   be able to measure a person's height
   be able to read a person's weight from scales marked in kilograms

> a. 0.39 means $\frac{3}{10} + \frac{9}{100} = \frac{39}{100}$
> b. Accurate measurement is essential in many school subjects, eg physics, woodwork, domestic science, chemistry etc. It is also essential in everyday life, eg measuring a room.

a. What is the length of this line?

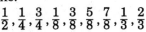

   Solution   *4.3cm*
b. Measure your height in metres (correct to the nearest 0.01m) and your weight in kilograms (correct to the nearest 0.1kg).

> Basic knowledge and understanding of the common fractions in daily use.

Understand the meaning of the following fractions in everyday life:
$$\frac{1}{2}, \frac{1}{4}, \frac{3}{4}, \frac{1}{8}, \frac{3}{8}, \frac{5}{8}, \frac{7}{8}, \frac{1}{3}, \frac{2}{3}$$
   Also tenths eg $\frac{3}{10}$
   And hundredths eg $\frac{7}{100}, \frac{17}{100}$.

> You should know that $\frac{3}{4}$ means 3 parts out of every 4 parts.

Level 4 – Number

Shade $\frac{5}{8}$ of this shape

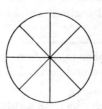

Solution   Any 5 sections

Basic use of percentages.

a. Know that % means out of 100.
b. Know that 10% of £18 means 18 pence out of every pound.

a. 27% means 27 out of 100 ($\frac{27}{100}$).
b. Decimal money (as opposed to the old pounds, shillings and pence) has encouraged the expression of quantities as percentages rather than as fractions. The metric system for length, weight (mass) and capacity has had the same effect.

a. Express 12 out of 50 as a percentage.
    Solution   24%
b. A meal costs £16 + 17.5% VAT. Calculate the VAT.
    Solution $(16 \times 17.5p) = 280p$ or £2.80

Place value of numbers.

Be able to express numbers in a variety of forms, eg
     4375 = 4 thousands + 3 hundreds + 7 tens + 5 units
     300 = 3 hundreds = 30 tens = 300 units

You need to be confident about expressing numbers in a variety of ways.

What is the value of the 4 in 3248?
    Solution   4 tens or 40 units

Level 4 – Number

> Multiplication tables and their use in everyday calculations and problems.

a. Possess instant recall of the 2, 3, 4, 5, 6, 7, 8, 9 and 10 times table.
b. Have the ability to make use of multiplication tables to perform multiplication and division problems.

> a. All tables as far as 10 × 10 must be memorised. You will be expected to perform simple multiplication and division questions mentally.
> b. Knowledge of multiplication tables speeds up many calculations.

a. Calculate mentally the cost of 6 packets of sweets at 8p each.
   Solution   48p
b. £54 is divided equally between nine girls. How much does each receive?
   Solution   £6

> Addition, subtraction, multiplication and division of whole numbers. This should be carried out mentally in the case of simple questions. Slightly more difficult questions should be carried out using paper and pencil methods. (A calculator would be allowed when multiplying a 3 digit number eg 728 × 426 or dividing by a 2 digit number eg 793 ÷ 13.)

Be proficient in addition, subtraction, multiplication and division of small numbers without the use of a calculator (pen and paper may be used).

> It is not always practical to use a calculator. If only a few sums are being performed it is much quicker to work them out in your head or to use a pen and paper. Try to devise your own method of performing mental calculations. (Basically – if the method works, use it). It is essential that you can do simple sums without the use of a calculator.

John ran 825 metres. Jayne ran 187 metres more than John. How far did Jayne run?
Solution   1 012 metres.

Level 4 — Number

Solution of mathematical questions which are written in the form of problems. (In the case of addition and subtraction up to 2 decimal places may be used.)

a. Be able to add or subtract decimals (including pounds and pence sums).
b. Be capable of short multiplication (eg 378 × 4) and short division (eg 739 ÷ 3) using pen and paper methods.
c. Know the mathematical techniques required to solve a problem.

You should be able to perform traditional addition and subtraction problems with and without a calculator.

a. A piece of wood 5m long is cut into two pieces. The longer piece is 3.87m. What is the length of the shorter piece?
   Solution   *1.13m.*
b. Stamps cost 14p. How many letters can be posted for £3 and how much change is there?
   Solution   *21 letters, 6p change.*
c. A bus can hold 73 passengers. There are 47 passengers on the bus. How many empty seats are there?
   Solution   *26.*

Check that the answers to mathematical questions are sensible. Numbers in the question should be rounded to allow an estimation of the answer.

Be able to round numbers, eg 721 is about 700, 888 is about 900. This allows you to make a reasonable guess at the answer to a calculation, eg 721 + 888 is about 700 + 900 = 1 600. Therefore the answer will be close to 1 600.

When using a calculator it is very easy to press the wrong key. The ability to estimate the answer mentally will help you to recognise a wrong answer.

Which of the following answers is the best guess for
                    9 121 − 3 932?
(1) 5 000   (2) 6 000   (3) 7 000   (4) 12 000.
Solution   *(1) 5 000.*

Level 4 – Number

> Write numbers which are given in decimal form correct to the nearest whole number.

Know that 8.6666666 on a calculator, written to the nearest whole number is 9.

> When calculating an answer an exact solution is not always required. For example, it would be ridiculous to give the distance between London and Manchester as 323.8762914. We need an approximate answer – 324 km is near enough.

When 88 is divided by 3 the calculator shows 29.3333333. What is the answer correct to the nearest whole number?
Solution 29.

> Calculators sometimes have rounding errors. This should be recognised and corrected.

Know that if a calculator display shows 1.99999999, then the correct answer is 2.

> Using some calculators, the sum 4 ÷ 3 × 3 produces an answer of 3.99999999. This is obviously incorrect. It contains a rounding error caused by the calculator working in decimals. The answer 3.99999999 should be corrected to 4.

Use your calculator to work out the answer to this question. Write down the answer given by your calculator and the correct answer. 797 ÷ 6 × 3.
Solution *Some calculators will give 398.499999. The correct answer is 398.5.*

> Use metric units of length, weight and capacity and understand the connection between units of different sizes eg 1 metre represents 100 centimetres.

Be able to measure length in millimetres, centimetres, metres and kilometres; weight in grams, kilograms and tonnes; capacity in millilitres, centilitres and litres.

Level 4 – Number

You need to know how to measure using different units and recognise the relationship between measurement in different units, eg if a box weighs 2.4 kg, this is the same as 2 400 g.

What is the length of this line

in

a. centimetres,

b. millimetres?

Solution   a. 2.4 cm   b. 24 mm.

Guess the length, weight and capacity of familiar everyday objects with a reasonable level of accuracy.

a. Know roughly what length, weight or capacity each of the following measures equal – millimetre, centimetre, metre, kilometre, gram, kilogram, tonne, millilitre, centilitre, litre, inch, foot, yard, mile, ounce, pound, stone, pint, gallon. Eg you should be able to indicate the length of a metre.

b. Know your height and weight in metric and Imperial units. (Imperial units are feet, inches, yards, miles, ounces, pounds, stones, pints, gallons.)

c. Know that a second is a very short space of time, know approximately how long a minute is, know the relationship between hours, minutes and seconds.
ie   60 seconds = 1 minute
       60 minutes = 1 hour
       24 hours = 1 day

d. Be able to add and subtract time.

Level 4 – Number

a. You should be able to use your knowledge of measures to make sensible estimates of length, weight and capacity, eg the length of a car.
b. Estimation is very useful in determining the accuracy of everyday measures, eg decide if a table will fit into a room.
c. You should understand and know how to use different units of time.
d. Time questions are common in examinations. A common error is to assume that there are 100 minutes in one hour. *Note:* you are advised not to use calculators when working out time questions.

a. What is the approximate length of the room you are in?

   Solution   *First guess and then measure to determine the accuracy of your guess.*

b. Find your own height in metric and Imperial units.

c. How many seconds are there in one day?

   Solution $60 \times 60 \times 24 = 86\,400.$

d. A train started its journey at 21.07 and arrived at its destination at 23.02. How long did the journey take in (i) minutes (ii) hours and minutes?

   Solution   *(i) 115 minutes (ii) 1 hour 55 minutes.*

Level 4 – Number

Attainment Test 4/1

Answers on page 141

1. Place these numbers in descending order.

    438, 192, 739, 804, 1 001.

2. Write these numbers in figures:
    (a) Three hundred and four
    (b) Half a million
    (c) Five thousand and twenty seven
    (d) Forty two thousand and fifty
    (e) One hundred and seven thousand.

3. Write these numbers in words:
    (a) 111   (b) 3 284   (c) 60 020.

4. Take twenty five thousand from a quarter of a million.

5. Write these numbers in ascending order:

    3 204, 893, 1 002, 638.

6. Write the answers to these multiplication sums:
    (a) 32 × 10 (b) 276 × 100 (c) 452 × 10 (d) 38 × 100 (e) 70 × 100.

7. What is the cost of 100 bars of chocolate at 17p each?

8. These scales show Mr White's weight in kilograms. What is his weight?

9. These milk cartons hold exactly one litre when full. Estimate the content of each carton. Choose from these quantities $\frac{1}{2}$ full, $\frac{1}{3}$ full, $\frac{2}{3}$ full, $\frac{1}{4}$ full, $\frac{3}{4}$ full.

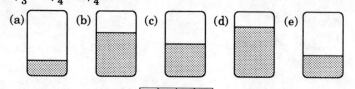

10. Shade $\frac{5}{8}$ of this shape.

11. Write these amounts as percentages:
    (a) 3 out of 100   (b) 4 out of 10   (c) 170 out of 1 000.

12. John has 100 sweets. 16% are toffees. How many toffees does he have?

13. Write 4 out of 25 as a percentage.

14. What is 12% of £14?

15. A meal cost £30 + VAT at $17\frac{1}{2}$%. How much is the VAT?

Level 4 – Number

Attainment Test 4/2 *Answers on page 141*

1. 800 people wished to tour a cave. They were split into groups of ten. How many groups were there?
2. What is the value of 35 hundreds?
3. David bought ten 19p stamps. How much change should he receive if he paid with a £5 note?
4. Carolyn bought the following items: butter 58p; cheese £1.26; loaf of bread 47p. What is the total cost?
5. What is the value of 5 in the number 3 529?
6. Jayne bought six bars of chocolate at 22p each. How much change did she receive from £2?
7. How many boxes of matches at 8p each can be bought with a £5 note, and how much change will there be?

Questions 8–10 *Choose which of the given answers is the best 'guess' for the answers to the questions.*

8. 3 896 + 2 901
   (a) 6 000
   (b) 6 700
   (c) 6 800
   (d) 6 900
9. 399 × 40
   (a) 1 600
   (b) 15 000
   (c) 16 000
   (d) 160 000
10. 3 703 − 2 799
    (a) 6 400
    (b) 1 100
    (c) 1 000
    (d) 900
11. John's calculator showed the following answers to some calculations. Write each answer correct to the nearest whole number.
    (a) 3.2 (b) 4.72934 (c) 5.5555555 (d) 0.8
12. Sarah used her calculator to solve this problem
    20 ÷ 3 × 6 ÷ 2.
    Her calculator showed the answer 19.999999. How should she write this answer?

Level 4 – Number

Attainment Test 4/3 *Answers on page 142*

1. Write 78 mm in centimetres.
2. Write 8.7 kg in grams.
3. Add 0.272 km and 387 cm. Write your answer in metres.
4. How many seconds are there in 8 hours?
5. A clock ticked every second. It was started at 9 am on 1st January 1992. At exactly what time would it reach one million ticks?
6. A sightseeing cruise is started at 11.27 am. The ship arrived back in port 5 hours 54 minutes later. What time did the ship arrive back in port?
7. A train travelled from Glasgow to London. It left Glasgow at 21.38 on Wednesday and arrived in London at 03.53 on Thursday. How long did the journey take?
8. Measure the length of these lines in millimetres.
   (a) ──────   (b) ──────────
9. Measure the length of the lines in question 8. Give your answer in centimetres.

Level 5 – Number

> Understand and use index notation for whole numbers.

a. Recognise that 3^4 means $3 \times 3 \times 3 \times 3$ and **not** 4×3.
b. Know that $3^4 \times 3^3 = 3^7$. (If the sign is multiply the powers are added.)
c. Know that $3^7 \div 3^4 = 3^3$. (If the sign is divide the powers are subtracted.)

> Index notation means using indices or powers (the '4' in 3^4). This provides a useful shorthand method of expressing mathematics. It is clearly quicker to write 8^6 than $8 \times 8 \times 8 \times 8 \times 8 \times 8$. Index notation is vital for the next topic **Algebra**.

a. What is the value of 4^3?
   Solution   64.
b. Express $7^3 \times 7^5$ as a single power of 7.
   Solution 7^8.
c. What is the value of $2^5 \div 2^2$?
   Solution $2^3 = 8$.

> Basic understanding of ratios.

a. Understand that ratio is a way of expressing one quantity as a proportion of another.
b. Understand the instruction 'Mix undiluted orange and water in the ratio 1 : 4 to make orange squash'.
c. Be able to use simple ratio in practical situations.

> We frequently use ratio in every day life without realising. If we say that an object is three times the size of another we mean a ratio of 3 : 1. The instruction on a bottle of orange squash may suggest a ratio of 1 : 4 meaning 1 part diluted orange to 4 parts water, ie you need four times as much water as undiluted orange. Know that the scale of a map or diagram can be expressed as a ratio, eg a scale in which 1 cm represents 1 m can be expressed as a ratio 1 : 100.

Write this ratio in its simplest form

8 : 20

Solution   2 : 5

Level 5 – Number

Carry out more difficult multiplication and division calculations without the use of a calculator.

a. Be able to work out 738 × 47 by a non-calculator method.
b. Be able to work out 638 ÷ 37 by a non-calculator method
(Pen and paper are allowed – you will be allowed to use any method that works.)

a. The important point is that you should have a method for calculating this type of question, be it traditional long multiplication or long division or some other method.
b. In the case of more difficult questions eg 5 378 × 439, 87 946 ÷ 37, etc calculators should be used. Pen and paper methods are no longer required.

a. 638 × 29
   Solution   18 502.
b. 827 ÷ 14
   Solution   59 remainder 1.

Work out simple fractions and percentages of quantities. A calculator may be used.

a. Know common equivalence between fractions and percentages. $\frac{1}{2} = 50\%$, $\frac{1}{4} = 25\%$, $\frac{3}{4} = 75\%$, $\frac{1}{3} = 33\frac{1}{3}\%$, $\frac{2}{3} = 66\frac{2}{3}\%$, $\frac{1}{10} = 10\%$, $\frac{1}{100} = 1\%$.
b. Know that 'find 3/8 of 20' means 3/8 × 20 and be able to calculate the answer. (***Answer** 7.5*)
c. Know how to use the % key on the calculator.
d. Know that percentage means 'out of 100'.
e. Know short-cut methods for calculating percentages.

Level 5 – Number

> a. You are expected to take short-cut methods when they are available, eg recognise that to find $\frac{1}{3}$ you divide by 3. To find $\frac{3}{4}$, you divide by 4 and then multiply by 3. When more difficult fractions are involved it is wiser to use a calculator.
> b. You should know that 'of' means 'multiply'.
> c. It is essential that you know how to calculate percentages because they are used frequently in examinations and every day life.
> d. 25% equals $\frac{1}{4}$. Therefore 25% of 20 means $\frac{1}{4}$ of 20 = 5.

a. Find $\frac{2}{3}$ of 12.
   Solution   8.
b. A meal costs £26 + 17.5% VAT. What is the total cost of the meal?
   Solution   £30.55
c. Write 21% as a fraction.
   Solution $\frac{21}{100}$
d. Find 75% of 40.
   Solution   30.

> Multiply and divide whole numbers with only 1 digit and a varying number of noughts mentally.

a. Recognise the similarity between 7 × 3 and 70 × 300 and use this similarity to work out the answer mentally.
b. Possess the same ability with division eg 6 000 ÷ 20 = 300.

> a. You must recognise that multiplying by 300 is the same as multiplying by 3 and adding two noughts. Multiplying by 70 is the same as multiplying by 7 and adding one nought.
> b. Dividing by 500 is the same as dividing by 5 and taking off two noughts. Dividing by 40 is the same as dividing by 4 and taking off one nought.

a. What is the area of a rectangle 30 m by 50 m?
   Solution   1 500 m².
b. 8 000 ÷ 200
   Solution   40.

Level 5 — Number

Calculate everyday problems involving negative numbers, eg temperature.

Understand the use of negative numbers in a real situation.

Now that we use the Celsius scale, temperatures in Britain in winter are often negative, therefore you need to understand that if the temperature is –8°C and there is a rise of 5°C, then the new temperature is –3°C.

How much higher is a man standing at the top of Mount Everest, height 29 000 feet, than a man floating in the Dead Sea which is 1 400 feet below sea level.
Solution   *30 400 feet.*

Use trial and improvement methods to solve problems.

Find the solution to problems by guessing the answer, testing that guess and then making a better guess.

To find the square root of 2.		
Try 1.5	$1.5^2 = 2.25$	Too high.
Try 1.3	$1.3^2 = 1.69$	Too low.
Try 1.4	$1.4^2 = 1.96$	Too low.
Try 1.42		
etc.		

Find the square root of 3 by 'trial and improvement'.
Solution   *1.73 Approx.*

Level 5 – Number

> Understand the meaning of significant figures and decimal places when rounding numbers to a given level of accuracy.

a. Approximate the answer to a question, ie give an answer which is about right but not exact.
b. Know that decimal places begin from the decimal point.

    7.281 correct to 2 decimal places is 7.28

    3.947 correct to 2 decimal places is 3.95

c. Know that significant numbers begin from the first number, eg

    3768 correct to 2 significant figures is 3800

    4283.12 correct to 3 significant figures is 4280

    3.023 correct to 2 significant figures is 3.0

    0.02798 correct to 1 significant figure is 0.03.

> Examination questions often ask for an answer to a specific level of accuracy, eg 3 decimal places or 2 significant figures. If this level of accuracy is not given exactly, marks will be lost.

a. 23 684 people attended a football match. What is this to the nearest hundred?
   Solution   *23 700.*
b. A train travels 510 km in 7 hours. What is its average speed in km/h. Give the answer correct to 2 decimal places.
   Solution   *72.86 km/h.*
c. 874 × 693. Give the answer correct to two significant figures.
   Solution   *610 000.*

> Understand scale and draw and interpret scale diagrams.

a. Draw scale diagrams.
b. Use scale to convert distances on a diagram or map into actual distances and vice versa.

Level 5 – Number

a. This is an important mathematical skill which is used in many school subjects such as science, geography and technical subjects.
b. Use scale to convert distances on a diagram or map into actual distances and vice versa.
c. Recognise that scale is an application of ratio in a practical situation.

a. i. Draw a plan of a room with a scale of 2 cm represents 1 m.
   ii. Use a 'kitchen planner' to draw a scale diagram of a kitchen. ('Kitchen planners' can be found in brochures for new kitchens.)

b. Use the scale on a road atlas to calculate the distance from Bristol to London. (*Note:* if the scale is given in the form 1 : 100 000, you would be expected to calculate that this is the same as 1 cm : 1 km.)
   Solution   *120 miles or 190 km (approx).*
c. A model ship is built to a scale of 1 : 500. The length of the full sized ship is 230 m. What is the length of the model ship? Give your answer in centimetres.
   Solution   *46 cm.*

Use Imperial units which are still used in everyday life. Know the approximate value of Imperial units in terms of metric units eg 1 gallon = 4.5 litres.

a. Know these Imperial units:

12 inches = 1 foot
3 feet = 1 yard
1 760 yards = 1 mile

16 ounces = 1 pound
14 pounds = 1 stone

8 pints = 1 gallon

a. Imperial units are still in regular use. You must be able to measure in both Imperial and metric units and be able to make sensible estimates of distances and weights without measures.
b. A dual system of measurement, metric and Imperial units, is used in Britain. It is therefore important for you to know the rough equivalents between the two systems.

Level 5 – Number

a. i. Convert 32 inches into feet and inches.
   Solution   2 feet 8 inches.
   ii. How many ounces are there in half a pound?
   Solution   8 ounces.
b. Petrol costs about £2.25 per gallon. What is the approximate price of petrol per litre.
   Solution   £0.50 per litre (approx).

Convert from one metric unit to another.

a. Memorise:

10 mm = 1 cm
100 cm = 1 m
1 000 m = 1 km

1 000 g = 1 kg
1 000 kg = 1 tonne

100 ml = 1 litre
1 000 cl = 1 litre

b. Be able to select an appropriate unit of measure for a variety of situations.

a. The metric system will eventually take over from the Imperial system. You must be fluent in the use of all listed metric measures.
b. You should appreciate that centimetres would be an inappropriate unit to measure the distance from London to Manchester. The distance should be measured in kilometres.

a. i. Know that a plug with a diameter of 5.3 cm will fit a plug hole of diameter 53 mm.
   ii. Re-write 0.32 kilograms in grams.
   Solution   320 grams.
b. Choose an appropriate metric unit to measure length of a pencil.
   Solution   Centimetres.

Level 5 – Number

Attainment Test 5/1
Answers on page 145

1. Express $7 \times 7 \times 7 \times 7$ using index notation.
2. What is the value of 6^3?
3. Which is the greater and by how much 5^3 or 3^5?
4. What is the value of $4^7 \div 4^5$?
5. Draw a scale diagram of this room, using a scale of 1 cm represents 1 m.
   What is the length of AC on your scale diagram?

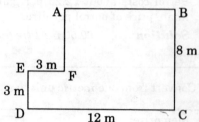

6. This is a scale drawing of a field. The distance from A to B is 80 metres.
   (a) What is the scale of the drawing?
   (b) How wide is the field? (ie from B to C)

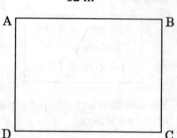

7. The scale of a plan of a room is 1 : 40.
   What length on the plan would represent 8 m?
8. A dress, normal price £25, was reduced by 10%. How much was the reduction?
9. A full box of chocolates contains 48 chocolates. Jennifer ate a quarter of them.
   (a) What percentage did Jennifer eat?
   (b) How many chocolates were left in the box?
10. A length of wood 10.5 metres long was cut into three equal pieces. How long was each piece.
11. Calculate 45% of £56.
12. Express 30 cm as a fraction of 1 m. Give your answer in its lowest terms.
13. John travelled 560 metres to school. He ran $\frac{3}{8}$ of the distance. How far did he run?
14. Express 400 m as a percentage of 1 km.

Level 5 – Number

Attainment Test 5/2   Answers on page 145

1. (a) What is the temperature shown on this thermometer?

   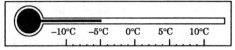

   (b) What is the temperature after a rise of 7°C?

2. The temperature at 2 pm was 3°C. At 8 pm the temperature was −5°C. Which of these statements describes accurately what has happened to the temperature?
   (a) The temperature has risen 5°C.
   (b) The temperature has fallen 5°C.
   (c) The temperature has fallen 3°C.
   (d) The temperature has fallen 8°C.

3. A submarine is 8000 metres below a plane which is flying at a height of 3500 metres above sea level. How many metres below the surface of the sea is the submarine?

4. (a) Write 742 correct to two significant figures.
   (b) Write 8937 correct to one significant figures.
   (c) Write 3.829 correct to two significant figures.
   (d) Write 0.372 correct to one significant figures.
   (e) Write 36.849 correct to two decimal places.
   (f) Write 7.908 correct to one decimal place.
   (g) Write 7.908 correct to two decimal places.

5. Use a trial and improvement method to find the square root of 27 correct to three decimal places. Show all your working clearly.

6. Calculate 18% of £1.27. Give your answer to the nearest penny.

7. An examination question read as follows:

   'Write ☐ correct to 2 decimal places'.

   Which of these numbers could be inserted into the box to make the answer to the question 3.83?
   (a) 3.84   (b) 3.837   (c) 3.8287   (d) 3.8733.

8. Add 350 cm and 3.1 m. Give your answer in metres.

9. Miss Jones' car weighs 945 kg. When Miss Jones' weight is added it makes exactly 1 tonne. What is Miss Jones' weight?

10. What is the total length of these pieces of wood?

    | 3m 28cm | 2m 6cm | 0.8m |

11. The size of my car engine is 1.6 litres. What is this in cubic centimetres?

12. A packet of coffee contains 750 g. How much coffee would I have altogether if I bought three similar packets?

Level 5 – Number

Attainment Test 5/3 *Answers on page 146*

1. A man is 5 feet 9 inches tall. What is this in inches?
2. What is the weight of a 1 kg bag of sugar to the nearest pound (lb)?
3. A car can travel 40 miles on one gallon of petrol. Write this distance in kilometres.
4. Mr Green wanted to buy a pint of milk when he was on holiday in Austria. The only cartons of milk he could find were 0.3 litres, 0.5 litres, 1 litre and 2 litres. Which carton is nearest to one pint?
5. The weight of American boxers is given in pounds. If a boxer weighed 225 lb, what would his weight be in stones and pounds?
6. Using a scale of 2 cm represents 1 m draw an accurate plan of this room.

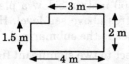

7. Look at this map. The actual distance from Ayden to Belfairs is 5 km.

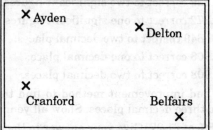

   (a) How many centimetres represent one kilometre?
   (b) What is the actual distance from Cranford to Delton?
8. This is an accurate scale diagram of a rectangular field ABCD. The scale is 1 cm represents 20 m.

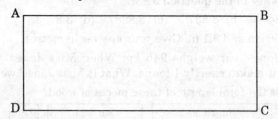

   (a) What is the actual length of the field?
   (b) What is the length from A to C in the diagram?
   (c) What is the actual distance from A to C in the field?
   (d) What is the area of the field?

Level 6 – Number

> Reading, writing and understanding the meaning of decimals.

a. Understand place value eg 0.345 means

	tenths	hundredths	thousandths
0 .	3	4	5

b. Know that moving the decimal point has the effect of multiplying or dividing by a factor of 10, ie:

to multiply by 10, move the decimal point one place to the right; to multiply by 100, move the decimal point two places to the right; to divide by 10, move the decimal point one place to the left.

eg 23.8 ÷ 100 = 0.238.

c. Be able to write decimals in ascending and descending order.

> a. You should understand that $\frac{3}{10}$ is larger than $\frac{27}{100}$.
> b. You need to understand the size of numbers, eg you should know that 0.4 is ten times larger than 0.04.
> c. You need to be able to recognise the smallest and largest numbers from a group of numbers. When presenting information you should be able to show it in a logical way, either in ascending order or descending order.

a. What is the value of the 7 in 5.673?

Solution   7 hundredths.

b. i. 6.3 × 100

Solution   630.

ii. 1.2 ÷ 10

Solution   0.12.

c. Write these numbers in ascending order.

0.09, 0.0875, 0.109, 0.096.

Method   Re-write the numbers to 4 figures after the decimal point:

0.0900

0.0875

0.1090

0.0960

Solution   0.0875, 0.09, 0.096, 0.109.

Level 6 – Number

Understand the connection between fractions, ratios, decimals and percentages.

a. Have a thorough understanding of the relationships between decimals and percentages.
b. Know that decimals and fractions are related,
eg $0.3 = \frac{3}{10}$, $0.37 = \frac{37}{100}$, $0.07 = \frac{7}{100}$.
c. Know that a fraction can be expressed in different but equivalent forms.
d. Know that a ratio can be expressed in different but equivalent forms.
e. Know the common relationships between fractions, decimals and percentages.

*Fraction	Decimal	Percentage	*Fraction	Decimal	Percentage
$\frac{1}{2}$ =	0.5	= 50%	$\frac{7}{8}$ =	0.875	= 87.5%
$\frac{1}{4}$ =	0.25	= 25%	$\frac{1}{3}$ =	0.333	= 33.3%
$\frac{3}{4}$ =	0.75	= 75%	$\frac{2}{3}$ =	0.667	= 66.7%
$\frac{1}{8}$ =	0.125	= 12.5%	$\frac{1}{10}$ =	0.1	= 10%
$\frac{3}{8}$ =	0.375	= 37.5%	$\frac{1}{100}$ =	0.01	= 1%
$\frac{5}{8}$ =	0.625	= 62.5%			

* This table should be memorised.

a. Fractions, decimals, percentages and ratios can all be used to express mathematics. You should be confident in making statements in any of the above forms.
b. Sometimes it is better to work with fractions than decimals, eg $\frac{1}{3}$ is often easier to use than 0.33333.
c. $\frac{1}{2} = \frac{2}{4} = \frac{3}{6} = \frac{4}{8}$ etc. ($\frac{1}{2}$ is the simplest form.)
d. $1 : 4 = 2 : 8 = 3 : 12$ etc. (1 : 4 is the simplest form.)

a. i. Write 27% as a decimal.
    Solution   0.27.
  ii. Write $\frac{3}{100}$ as a percentage.
    Solution   3%.

Level 6 – Number

b. Write 0.72 as a fraction in its lowest terms.
   Solution $\frac{18}{25}$
c. Fill in the missing numbers $\frac{3}{4} = \frac{?}{12} = \frac{15}{?}$.
   Solution $\frac{3}{4} = \frac{9}{12} = \frac{15}{20}$.
d. Write the ratio 20 : 15 in its simplest form.
   Solution 4 : 3.

> Increase and decrease quantities by given percentages and fractions.

a. Know how to use percentages to work out profit and loss, increases and decreases, etc.
b. Be able to reduce or increase amounts by simple fractions. Fractions will normally have simple denominators eg halves, thirds, quarters, fifths, eighths, tenths and hundredths.

a. You should be able to work out price rises, price reductions, interest rates, VAT, profit and loss, wage rises, etc.
b. You should be happy to increase and decrease by percentages or fractions, eg the price of a dress is reduced by one quarter.

a. A man earning £185 receives a 5% pay rise. What is his new wage?
   Solution £194.25.
b. i. What is $\frac{2}{3}$ of £75.
   Solution £50.
   ii. Reduce 72 by $\frac{3}{8}$.
   Solution 45.

> Use ratio in a variety of problems.

Know that ratio can have a wide variety of mathematical uses, eg showing the relationship between one quantity and another.

Level 6 – Number

We often use ratio without realising. If 8 books cost £12 and we wish to find the cost of 10 books we can use ratio. The price increases in the ratio 10 : 8 or 5 : 4.

Divide £100 in the ratio 3 : 2.
Solution   £60 : £40.

Convert from fractions to decimals and fractions to percentages, and have an elementary knowledge of percentages.

Be able to convert fractions to decimals and percentages. The common relationships should be memorised. (See Dog, page 24.) For other values calculators should be used.

Percentages are extremely important in examination work and in everyday life, eg a shop can advertise a 25% reduction on all marked prices. This saves them from changing all of the price tickets.

17 out of 25 people who were interviewed watched the 6 o'clock news on television. Express this as a percentage.

Solution   Method $\frac{17}{25} \times 100 = 68\%$.

Be able to check and make rough estimates of the answers to multiplication and division questions.

a. Be able to recognise a wrong answer, particularly an answer which is wrong by a factor of 10, ie a decimal point in the wrong place.
b. Be able to approximate numbers and then use the approximation to estimate an answer.

Level 6 – Number

> a. People sometimes give ridiculous answers to questions, eg calculating a personal quarterly phone bill of £4 825, when the answer should be £48.25.
>
> b. When using a calculator it is very easy to press a wrong key. You should always estimate the answer and check if your final answer is sensible.

a. Choose the most suitable estimate for the height of a classroom ceiling.
   (1) 3 cm   (2) 30 cm   (3) 3 m   (4) 30 m.
   Solution   3 m.

b. Estimate the cost of 3 268 kg of sugar at £0.95 per kg.
   Method   3 268 is about 3 000; £0.95 is about £1.
   Solution   3 000 × 1 = £3 000.

Level 6 – Number

Attainment Test 6/1
Answers on page 149

1. In the number 37.245 the value of the 2 is
   (a) 2 units (b) 2 tenths (c) 2 hundredths (d) 2 thousandths.
2. In the number 42.683 the value of the 3 is
   (a) 3 units (b) 3 tenths (c) 3 hundredths (d) 3 thousandths.
3. 0.23 is the same as
   (a) 23 units (b) 23 tenths (c) 23 hundredths (d) 23 thousandths.
4. Write 0.109, 0.098, 0.1103, 0.0783 in ascending order.
5. Fill in the missing numbers. (a) $\frac{3}{5} = \frac{?}{10}$ (b) $\frac{12}{16} = \frac{9}{?}$
6. Write 0.7 as a fraction.
7. Write $\frac{3}{8}$ as a decimal.
8. Write 0.2 as a percentage.
9. Write 0.385 as a percentage.
10. Write 0.675 as a fraction in its lowest terms.
11. Write $\frac{7}{16}$ as a percentage.
12. What is $\frac{5}{8}$ of £24?
13. Reduce 39 by $\frac{2}{3}$.
14. Increase 50 by one quarter.
15. In a sale a coat which normally cost £30 was reduced by £6. What percentage reduction was this?
16. A dress which normally cost £24 was reduced by 25%. What was the sale price of the dress?
17. A new car costs £16 400. Next month there will be an 8% price rise. What will be the cost of the car next month?
18. The normal price of a television is £300. In a sale there is a price reduction of 20%. How much money is saved by buying the television in the sale?

Attainment Test 6/2 *Answers on page 150*

1. Simplify these ratios.
   (a) 4 : 20 (b) 15 : 12 (c) 4 : 6 : 8 (d) $\frac{1}{2} : \frac{1}{4}$.
2. Sixty sweets are divided between Sarah and John in the ratio 3 : 2. How many sweets does Sarah receive?
3. In a will, money is left to three brothers Alan, Bob and Clive in the ratio 5 : 4 : 3. Bob received £2 000 more than Clive. How much money did Alan receive?
4. This is the recipe for Yorkshire Pudding for eight people:
   10 oz of flour, $1\frac{1}{2}$ pints of milk, 2 eggs.
   How much of each ingredient should you use for
   (a) 4 people (b) 20 people?
5. 27 people out of 50 watched the 9 o'clock news on television. What percentage is this?
6. A shopkeeper bought eggs at 60p a dozen. He sold them for 75p a dozen. Express his profit as a percentage of the cost price.
7. In a survey 85 people out of 500 stated that their favourite meal was fish. What percentage is this?
8. Express 75 as a percentage of 250.
9. 263 ÷ 37 is about
   (a) 6 (b) 7 (c) 8 (d) 9.
10. 58 × 31 is approximately equal to
    (a) 180 (b) 1 600 (c) 1 700 (d) 1 800.
11. Which of these answers is sensible for the height of a man?
    (a) 1.76 cm (b) 17.6 cm (c) 176 cm (d) 1 760 cm.
12. 25 books cost £40. What is the cost of 10 books?

Level 7 – Number

Write a whole number as a product of prime numbers.

a. Be able to break a whole number down into its prime factors.
b. Be able to find the HCF (Highest Common Factor) and LCM (Lowest Common Multiple) of two or more numbers.

There are various ways of finding the HCF and LCM, eg reducing the numbers to prime factors. You should know the prime numbers 2, 3, 5, 7, 11, ...

a. Express 756 as a product of primes.

Method keep dividing by prime numbers in ascending order.

```
2 )  756
2 )  378
3 )  189
3 )   63
3 )   21
7 )    7
       1
```

Solution $2 \times 2 \times 3 \times 3 \times 3 \times 7 = 2^2.3^3.7$

b. Find the HCF and LCM of 18 and 24.

Method HCF =   2 × 3   = 6
                                  ↑   ↑
             18 =   2 × 3 × 3
             24 =   2 × 2 × 2 × 3
                      ↓   ↓   ↓   ↓   ↓
             LCM =   2 × 2 × 2 × 3 × 3   = 72

Solution   72.

Multiply and divide single figure sums where the single figure has noughts in front or behind eg 300 × 0.02. Understand that multiplying by a number less than 1 produces a smaller number, dividing by a number less than 1 produces a larger number.

Level 7 – Number

a. Be able to work out mentally questions such as
    200×30, $800 \div 40$, 70×4, $9\,000 \div 30$, etc.

b. Be able to work out mentally questions such as the following, and know whether the answer will be larger or smaller than the question.
    30×0.3, $8\,000 \times 0.6$, $60 \div 0.05$, $80 \div 0.2$.

a. The ability to perform simple arithmetic calculations is a valuable skill, particularly when estimating – eg if we know that a car travels at a speed of 80 kilometres per hour we can estimate the time required for a journey of 400 kilometres.

b. Decimal working makes frequent use of this skill.

a. A school has 40 classes with 30 pupils in each class. How may pupils attend the school.
  Solution   1 200.

b. How many sacks containing 0.02 tonnes of coal can be filled from 4 tonnes?
  Solution   200.

Use a calculator for multiplication and division of numbers where the numbers are large.

Be able to use a calculator accurately.

Many problems in real life do not have round numbers, eg find the area of a rectangular field of length 382.48 metres and width 68.37 metres. A calculator is needed to answer these questions, although the approximate answer should be estimated. This will eliminate many careless errors.

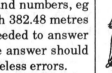

i. 2480 packing cases each containing 144 video cassettes were delivered to a shop. How many cassettes were delivered?
  Solution   357 120.

ii. £1 = \$1.6328. What is the value of \$1 120?
  Solution   £685.94 (approx.).

Level 7 – Number

Use some of the more sophisticated keys on a calculator such as the memory or brackets when performing more complex calculations.

a. Be able to make efficient use of the memory and brackets when using a calculator.
b. Recognise the difference between these two questions:

   (i) $\dfrac{3.8 - 2.6}{4}$ (ii) $3.8 - \dfrac{2.6}{4}$

(Solution (i) 0.3 (ii) 3.15.)

At this stage you should have an excellent knowledge of the way that the memory and brackets work. Different calculators work in different ways, therefore it is important to become familiar with one calculator and always use that one.

$$\dfrac{3.92 + 4.7}{3.6 - 1.14}$$

Solution   3.50407 approx.

Extension of the use of existing numerical skills in attempting more complex calculations.

Be able to make use of all types of mathematics in a wide range of questions and previously unknown situations, eg other school subjects such as physics.

You should be able to adapt your knowledge to fit unknown situations and previously unseen questions in mathematics or other subjects.

How long will a spaceship travelling from earth take to reach the moon at a speed of 2 500 metres per second? (The distance to the moon is a quarter of a million miles.)

Method   (1) 5 miles = 8 kilometres, therefore the distance to the moon is 400 000 kilometres.

(2) 2 500 metres = 2.5 kilometres.

(3) 1 hour = 3 600 seconds, therefore the spaceship travels 9 000 km in one hour.

(4) $\dfrac{400\ 000}{9\ 000}$

Solution   44.44 hours.

Level 7 – Number

Combine two linear measurements to form a compound measure – eg miles and hours are linear measures; miles per hour is a compound measure.

Understand and be able to calculate speed and density. The following formulae should be memorised.

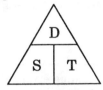
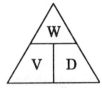

Speed = $\dfrac{\text{Distance}}{\text{Time}}$ Volume = $\dfrac{\text{Weight}}{\text{Density}}$

Time = $\dfrac{\text{Distance}}{\text{Speed}}$ Density = $\dfrac{\text{Weight}}{\text{Volume}}$

Distance = Speed × Time   Weight = Volume × Density.

Compound measures are units such as miles per hour, feet per second, (these are speed); g/cm³ (this is density). Compound units are formed by linking two units of measure, eg miles and hours.

The triangles above help us to remember the formulae

eg $S = \dfrac{D}{T}$, $T = \dfrac{D}{S}$, $D = S \times T$.

Find the speed of a car which travels 300 km in 5 hours.

Method   Look at the triangle. This tells us that:

Speed = $\dfrac{\text{Distance}}{\text{Time}}$

Speed = $\dfrac{300 \text{ km}}{5 \text{ hours}}$

Solution   *60 km/h.*

Know that measurement is not exact. It is only accurate within certain limits. Select appropriate units and levels of accuracy for measurement in different situations.

Know that accuracy is only needed to a certain degree.

Level 7 – Number

When measuring glass to fit a window you would need to know the size to the nearest millimetre. However, when measuring the playground you would not need this level of accuracy. The length to the nearest metre would probably be sufficient for most purposes.

What units would you choose to measure the length of a mouse?

Solution   Centimetres (or inches).

Know that measurement involves rounding errors of up to half a unit above or below the stated length, weight or capacity.

Know that measurements can contain rounding errors, eg 7.38 grams means that this is the weight to two decimal places. Had the measurement been taken to three decimal places the measurement could be anywhere between 7.375 and 7.385 grams.

If an object is weighed and the weight is stated to be 254 grams, this means that the weight is approximately 254 grams. If a length is given as 3.84 km the exact length is between 3.835 and 3.845 km.

The length of a desk was given as 175.3 cm. Give the maximum and minimum possible lengths of the desk.

Solution   Maximum length is 175.35 cm.
              Minimum length is 175.25 cm.

Level 7 – Number

Attainment Test 7/1 *Answers on page 153*

1. Express the following numbers as products of primes.
   (a) 225 (b) 300 (c) 504.
2. Find the HCF (Highest Common Factor) of 108 and 180.
3. Find the LCM (Lowest Common Multiple) of 140 and 160.
4. A shopkeeper has 80 boxes of apples, each box weighs 1 kg. The average weight of each apple is about 100 g. Approximately how many apples does the shopkeeper have?
5. Assume that £1 = 3.02 DM (German marks), calculate the following.
   (a) The number of pounds (£'s) you would receive for
      (i) 38 DM.
      (ii) 142 DM.
      (Give your answer to the nearest penny.)
   (b) The number of German marks you would receive for
      (i) £13.
      (ii) £28.34.
      (Give your answer correct to 2 decimal places.)
6. Given that 1 kg = 2.2 lb, convert
   (a) 5.9 lb to kg.
   (b) 7.92 kg to lb.
7. Use your calculator to evaluate these expressions.
   (a) $\dfrac{5.8 - 3.2}{6.7 + 2.4}$.
   (Give your answer correct to 3 significant figures.)
   (b) $5.8 + \dfrac{3.2}{6.7} + 2.4$.
   (Give your answer correct to 4 decimal places.)
8. Find the value of A, B and C in this sum.

   $$\begin{array}{r} 7\ 2\ A \\ -\ 2\ B\ 6 \\ \hline C\ 7\ 7 \end{array}$$

9. An aircraft covers 468 km in 1 hour 20 minutes.
   (a) What is its average speed in km/h?
   (b) What is its average speed in m/s (metres per second)?

Level 7 – Number

Attainment Test 7/2 *Answers on page 153*

1. A train travels 240 km in 3 hours 10 minutes. What is the speed in km/h:

   (a) Correct to two decimal places;

   (b) Correct to the nearest ten kilometres per hour.

2. An aircraft travels at a speed of 330 km/h for a distance of 154 km. How long does the journey take? Give your answer in minutes.

3. A block of metal is 0.5 m long, 30 cm wide and 20 cm high. The density of the metal is 15.2 g/cm^3. What is the weight of the block of metal?

4. What is the density (in g/cm^3) of a block of wood which weighs 34 kg and measures 50 cm by 40 cm by 20 cm?

5. A car starts its journey at 10.51 and arrives at its destination at 14.48. The distance covered is 242 km. What is the average speed of the car to the nearest kilometre per hour.

6. A man drives from his house to visit a friend who lives 240 km away at an average speed of 80 km/h. He returns at a speed of 60 km/h. What is his average speed for the whole journey?

7. Choose the most sensible degree of accuracy for the length of a pencil.

   (a) 36 mm   (b) 36.1 mm   (c) 36.08 mm   (d) 36.083 mm.

8. Which of these units is the most appropriate for measuring the distance from the Earth to the Moon?

   (a) Kilometres   (b) metres   (c) centimetres   (d) millimetres.

Algebra

Algebra is the theoretical part of mathematics. You will find that many algebraic questions are a challenge, and that they require you to use all your powers of deduction and logic. You will find the ability to perform algebraic calculations an important requirement for physics and chemistry.

Key

 The blackboard represents the level of attainment required.

 The dog holding a calculator shows the skills and knowledge required for this topic.

 The talking parrot gives you the author's comments.

? The question mark stands for typical tests and questions that you will have to complete.

Level 4 – Algebra

Identify number patterns in order to predict the next number in the sequence.

a. Recognise number patterns such as $3\,000 \times 4 = 300 \times 40 = 30 \times 400 = 3 \times 4\,000$.

b. Recognise the equivalence of fractions, eg

$$\frac{1}{6} = \frac{2}{12} = \frac{3}{18} = \frac{4}{24} \text{ etc}$$

a. This tests your ability to manipulate numbers. You will only acquire the necessary strategies by attempting numerous questions of various types. IQ tests provide good practice.

b. You should understand that a fraction can be expressed in a variety of equivalent forms.

a. Complete the next term in this series.

$$128 \times 1 = 64 \times 2 = 32 \times 4 =$$

Solution *16 × 8.*

b. What is the missing number?

$$\frac{?}{8} = \frac{3}{6}$$

Solution *4.*

Find mathematical rules to explain how to find the next number in a sequence.

a. Possess the ability to express mathematical concepts in words.

b. Possess the ability to examine mathematical data and reach a conclusion based upon your findings.

School mathematics is moving away from simply learning facts and mathematical techniques. You will now be asked to examine mathematical data and comment upon your findings. This ability is examined through coursework. The investigations can take a few minutes or several hours to complete. You must present a logical argument to support you findings. There is often a variety of correct solutions.

The best way to learn how to investigate is to try questions from mathematical puzzle books.

Level 4 – Algebra

Understand mathematics that is written in words instead of numbers. (This is an introduction to algebra where letters will be used instead of numbers or words.)

a. Be able to use a simple formula.
b. Be able to explain how two sets of data are mathematically related.

a. We often use simple formulae, eg the gas bill is calculated by this formula:

Gas Bill = Fixed Charge + cost of units used.

b. This is a skill which is only acquired by practice. The information has to be studied and a mathematical connection found.

a. This is the formula for hiring a car

Hire Charge = Fixed Charge + 5p per mile.

What is the cost of hiring a car if the fixed charge is £10 and the car travels 120 miles?

Solution £16.

b. Explain how these two sets of data are mathematically related.

7 ⟶ 22
8 ⟶ 25
9 ⟶ 28
10 ⟶ 31

Solution *Multiply by 3 and add 1.*

Division and multiplication have opposite effects. This can be used to check answers.

Know multiplication is the inverse or 'opposite' of division, division is the inverse or 'opposite' of multiplication.

This is an introduction to the manipulation of numbers in algebraic form (ie using letters instead of numbers).

		Using numbers
If	$a \times b = c$	$7 \times 8 = 56$
Then	$c \div a = b$	$56 \div 7 = 8$
And	$c \div b = a$	$56 \div 8 = 7.$

Level 4 – Algebra

i. $R = S \times T$. What does S equal?

Solution $S = R \div T$.

ii. The cost of a taxi fare is given by the formula

$$\text{FARE} = \text{Distance (in miles)} \times £2.$$

If the fare was £10, what distance was travelled?

Solution   *5 miles.*

Know that the co-ordinates (3,2) means 3 across and 2 up. This is only needed in the first quadrant.

a. Be able to give the co-ordinates of a point.
b. Be able to draw simple graphs.

a. *Note:* In the first quadrant (top right) all co-ordinates are positive.

This is the first quadrant

b. You should be able to draw graphs and understand co-ordinates.

a. What is the co-ordinate of the point marked with a cross?

Solution   *(1,3).*

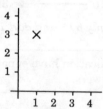

b. If your school has a computer, you should be capable of using LOGO commands on it.

Level 4 – Algebra

Attainment Test 4/4 Answers on page 142

1. This is a sequence.   3 × 4 = 12
   Fill in the missing numbers. 3 × ☐ = 15
   ☐ × 6 = ☐

2. What is the next term in this sequence?

   $\frac{1}{2} \times 20, \quad \frac{1}{4} \times 10, \quad \frac{1}{8} \times 5, \ldots\ldots$

3. The price of a meal is calculated using this formula:
   Price of Meal = Cost of Ingredients + Cost of Electricity.
   Find the cost of ingredients if the price of the meal is £7.28 and the cost of electricity is 74p.

4. The telephone bill is calculated by the following formula:
   Telephone Bill = Fixed Charge + 4.8p per unit used.
   (a) What is Mr Green's telephone bill if the fixed charge is £7.80 and 385 units are used?
   (b) How many units did Miss White use if her bill was £30.36 and the fixed charge was £7.80?

5. What is the rule which connects these two sets of data?

   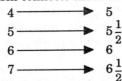

6. What is the value of y in each of these equations?
   (a) 3y = 21 (b) y + 6 = 8 (c) y − 8 = 3 (d) y + 10 = 4 (e) $\frac{y}{3} = 10$

7. If A = 20B, what does B equal in terms of A?

8. What are the co-ordinates of the mid-point of the line joining the points (1,4) and (5,1)?

9. This is a map of an island. The area of each square is 1 km².

   What is the approximate area of the island?

10. What is the next fraction in this sequence?

    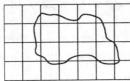

Level 5 – Algebra

Understand the meaning of factors, multiples, prime numbers, square numbers, cube numbers, square roots and cube roots.

a. Be able to write multiples of any number.
b. Be able to find the factors of numbers.
c. Know, or know how to find, all prime numbers up to 31, ie 2, 3, 5, 7, 11, 13, 17, 19, 23, 29, 31.
d. Know what a square number and a cube number are, know all of the square and cube numbers listed below.

Number	Square	Cube
1	1	1
2	4	8
3	9	27
4	16	64
5	25	125
6	36	216
7	49	343
8	64	512
9	81	729
10	100	1 000

e. Know what square roots and cube roots are and know all of the cube roots and square roots of the numbers in the table above.

a. *Example.* Multiples of 8 are 8, 16, 24, 32 . . . 80, 88, . . . etc, ie any number in the 8 times table.

b. *Example.* Factors of 24 are 1, 2, 3, 4, 6, 8, 12 and 24, ie numbers which divide into 24 exactly without a remainder

c. A prime number is a number which is only divisible by 1 and itself.

d. A square number is produced by multiplying a number by itself, eg $13^2 = 13 \times 13 = 169$. Therefore 169 is a square number.

A cube number is produced in this way:

$12^3 = 12 \times 12 \times 12 = 1\ 728$. Therefore 1 728 is a cube number.

e. The square root of 49 is 7 (or -7), because $7 \times 7 = 49$.

The cube root of 125 is 5, because $5 \times 5 \times 5 = 125$.

To find the square root use the $\sqrt{}$ key, to find the cube root use the $\sqrt[3]{}$ key on your calculator.

Level 5 – Algebra

5, 7, 12, 16, 85, 196, 216, 302.

From this list write down:

a. a multiple of 3;
b. a factor of 72;
c. a prime number;
d. i. a square number
   ii. a cube number;
e. i. the square root of 144
   ii. the cube root of 343.

Solutions   a. 12 or 216; b. 12; c. 5 or 7;
            d. i. 16 or 196 ii. 216;
            e. i. 12 (-12 is also a square root) ii. 7.

Recognise patterns which are presented in the form of diagrams or pictures.

a. Be able to recognise patterns. These are often of the IQ type. Important patterns are:

    •   ::   :::   ::::   square numbers

    •   .:   .::   .:::   triangular numbers

b. Be able to transfer the above methods to three dimensional shapes, eg cubes.

a. The IQ type question is designed to make you think. These questions are often the introduction to an investigation, which you complete as part of your coursework.
b. You should be able to understand two dimensional drawings of three dimensional shapes.

a. Find a relationship between square and triangular numbers.

   Square numbers   1   4   9   16   25   36
   Triangular numbers 1   3   6   10   15   21

   Solution   1   3   6   10   15

Difference between the square and the triangular numbers equals the triangular number sequence.

b. How many small cubes are required to make the next shape in the sequence?

   Solution   64.

Level 5 – Algebra

Understand and use simple computer programs.

a. Follow instructions which are given in the form of a computer program or flow chart diagram.
b. Be able to find the next terms in sequences such as 1, 1, 2, 3, 5, 8, 13, 21, ...

a. We are now in the computer age and instructions are often given as lists or flow diagrams. You may be asked to write a flow diagram containing instructions for making a cup of tea or use a mathematical program to produce the three times table, as shown below.
b. You should be able to find the next terms and explain how you obtained the answer, ie the rule which you have found.

a. Follow the instructions in this program:
```
10 for number = 1 to 8
20 print number + number + number
30 next number
40 end
```

Solution   3, 6, 9, 12, 15, 18, 21, 24.
(This is the 3 times table)

b. Write down the next two numbers in this sequence and give the rule.

1, 1, 2, 3, 5, 8, 13, 21, 34, 55, ...

Solution   89, 144. (Rule – add the two previous numbers.)

Understand mathematics which is presented in algebraic form, ie using letters to represent numbers.

a. Understand algebraic expressions such as
   $3a$ means $3 \times a$ a^3 means $a \times a \times a$,
   ab means $a \times b$ $3(a + b)$ means $3a + 3b$.
b. Be able to use a formula.

a. You should be confident about using mathematical expressions which contain letters in place of numbers.
b. At Level 4 you were asked to substitute numbers for formulae and equations expressed in words. At Level 5 letters are used instead of words.

Level 5 – Algebra

a. What is the value of $4a + b^3$, when $a = 5$ and $b = 4$?
   Solution   84.
b. Use the formulae below to find the area of a trapezium.
   $\frac{1}{2}(a + b) \times h$, when $a = 3$, $b = 7$ and $h = 4$.
   Solution $\frac{1}{2}(3 + 7) \times 4 = 20$.

> Use letters to represent numbers.

Be able to express a mathematical statement which is given in algebraic form.

> You need to express mathematical statements using algebra, ie letters instead of numbers. This will help you to produce a formula which will work in all cases irrespective of the input numbers.

A man buys X first class stamps at 20p each and Y second class stamps at 16p each.
i. Express this statement algebraically.
ii. Make an equation to show the total cost (£T).
Solution   i. $(20X + 16Y)$ pence.
            ii. $£T = £(0.20X + 0.16Y)$.

> Use co-ordinates to plot points in all four quadrants. This involves using negative numbers.

a. Know how to describe the location of any point in any of the four quadrants.

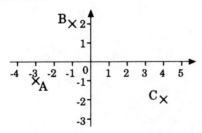

b. Know how to draw shapes and lines using all four quadrants.

Level 5 – Algebra

> a. Points are described using (x,y) notation: x is the value on the horizontal axis, y is the value of the vertical axis. The point marked A would be described as (–3,–1).
>
> b. Graphs are very important: you must understand this section to be able to answer questions in the section on **Handling Data**.

a. What are the co-ordinates of the points marked B and C on the graph at the bottom of page 45.
Solution   B (–1,2) C (4,–2).

b. Draw the quadrilateral with the co-ordinates (1,2), (2,–2), (–3,–2), (–1,2). What is the mathematical name of this shape?

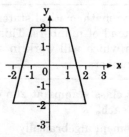

Solution   A trapezium.

Level 5 – Algebra

Attainment Test 5/4   *Answers on page 146*

1. List all of the prime numbers between 10 and 20.
2. What is the next prime number after 23?
3. Look at this list of numbers.
   $$5, 8, 12, 18, 21.$$
   (a) Write down a multiple of 7.
   (b) Write down a factor of 16.
4. Complete the next term in these sequences.
   (a) 1, 4, 9, 16, 25, 36, ...
   (b) 1, 1, 2, 3, 5, 8, 13, 21, ...
   (c) 1, 8, 27, 64, 125, ...
   (d) 1, 2, 4, 7, 11, 16, 22, ...
5. What are the solutions to these questions?
   (a) $\sqrt{64} - \sqrt[3]{64}$ (b) $5^3 - 4^3$ (c) $6^3 - \sqrt[3]{125}$.
6. Complete the next set of dots in each of these patterns.
   (a)
   (b)
   (c)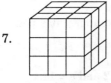

7. How many small cubes are needed to make this large cube?

8. Use the flowchart to find the smallest square number which is greater than 500. Start with N = 15.

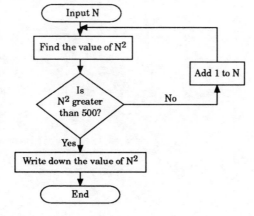

Level 5 – Algebra

Attainment Test 5/5
Answers on page 147

1. Given a = 1, b = 2, c = 3, d = 4 and e = 0, find the values of
   (a) ab (b) 3c (c) 2c – d (d) 3cd (e) cde (f) 4d – 3e.

2. The area of a triangle is calculated by using the formula
   $$\text{Area} = \frac{1}{2}(b \times h).$$
   What is the area of a triangle where b = 8 m and h = 10 m? (Give your answer in m².)

3. (a) Give the co-ordinates of the point p.
   (b) On the graph plot these points
   A (2,2), B(3,–2), C(–1,–3), D (–2,1).
   Join the points to form a quadrilateral. What is the mathematical name of this quadrilateral?

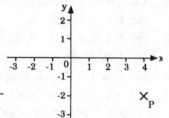

4. John buys x bars of chocolate at 17p each and y packets of sweets at 14p each.
   (a) Express the statement mathematically.
   (b) Complete the following equation to show the total cost (£T) of the sweets and chocolate.

   £T = ☐

5. Miss White buys X apples at r pence and Y oranges at s pence each.
   (a) Express the statement mathematically.
   (b) Find the total cost when X = 4, Y = 3, r = 12 and s = 14.

Level 6 — Algebra

> Investigate and find a rule to produce the next terms in a sequence.

Investigate and find a rule to produce the next terms in a sequence.

> These are IQ type questions. Numbers in the sequence may be linked by addition, subtraction, multiplication, division, or some more complex method.

What is the next number in this sequence?

$$3, 4, 6, 9, 13, 18, \ldots$$

Solution   24.

> Use a computer to investigate the terms of a sequence and find a general rule for the sequence.

Be familiar with the use of computers. Be able to access data. Be able to interrogate the computer to obtain the required information.

> Use of the computer is now an integral part of mathematics.

Use a computer to find the numbers under 5 000 which are both square numbers and cube numbers.
Solution   1, 64, 729, 4 096.

> Use a variety of techniques to solve linear equations.

a. Know all the rules of equations, ie be able to manipulate equations, understand that the sign changes when numbers 'go through' the equals.

b. Understand how to solve linear equations.

Level 6 – Algebra

a. You need a method of solving equations. This may be by traditional methods, flow diagrams, etc. Although 'trial and error' or guesswork methods are acceptable, they often fail and can take considerably more time than the 'proper' methods.

b. A linear equation is an equation in x, eg $x + 3 = 5$. (*Note:* if x is written to a power, eg x^2, x^4, etc, it is not linear but a polynomial.)

a. i. Solve $x + 7 = 12$
   ii. Solve $x - 3 = 8$
   iii. Solve $3x = 18$
   iv. Solve $\frac{x}{3} = 15$
   Solutions   *i. 5   ii. 11   iii. 6   iv. 45.*

b. Solve $7x - 4 = 5x + 10$
   Solution   *x = 7.*

Use a trial and improvement method to solve more complex forms of equation where the unknown is expressed as a power eg $y^3 = 18$.

Be able to solve more difficult equations by guessing the value of x, testing the value and finding a better approximation.

You should be able to use a trial and improvement method, eg to solve $x^2 = 7$.

				Low	High
Guess 1	x = 2,	2×2	= 4	2	
Guess 2	x = 3,	3×3	= 9		3
Guess 3	x = 2.5,	2.5×2.5	= 6.25	2.5	
Guess 4	x = 2.75,	2.75×2.75	= 7.5625		2.75
Guess 5	x = 2.7,	2.7×2.7	= 7.29		2.7
etc.					

Use a trial and improvement method to find an approximate value for x given $x^3 = 12$.

Solution   *A method such as that shown above. Value of x is 2.289 approx.*

Level 6 – Algebra

> Draw a graph of a line or a function.

a. Be able to plot the graph of y = x + 5 (this can be written as x → x + 5).

b. Be able to plot the graph of y = 3x² (this can be written as x → 3x²).

> a. You should be able to complete a table of values for a linear function and then plot the function on a graph (ie draw the graph).
> b. You should be able to complete a table of values for a polynomial function and then draw the graph.

a. Complete the table of values for the graph of y = x – 3 and draw the graph.

x	–2	–1	0	1	2	3
y	–5		–3			0

Solution   Missing values –4, –2, –1.

Graph should be a straight line through these points (–2,–5) (–1,–4) (0,–3) (1,–2) (2,–1) (3,0).

b. Complete this table of values and draw the graph of y = 3x² (ie x → 3x²).

x	–3	–2	–1	0	1	2
y						

Solution   Values of y are 27, 12, 3, 0, 3, 12.

Graph should be a curve through these points (–3,27) (–2,12) (–1,3) (0,0) (1,3) (2,12).

Level 6 – Algebra

Attainment Test 6/3

Answers on page 150

1. What are the next two numbers in this sequence?

   9, 16, 35, 72, 133, 224, ...

2. What are the next two numbers in this sequence?

   7, 15, 39, 111, 327, 975, ...

3. Solve these equations.
   (a) $3x = 21$ (b) $5x - 7 = 3x + 12$ (c) $4x - 3 = 6x + 8$
   (d) $4(2x - 3) = 5x$ (e) $\frac{3x}{4} = 7.5$

4. Solve these equations using a trial and improvement method. Show all of your working and continue working until you have an answer correct to 3 decimal places.
   (a) $x^2 = 7$ (b) $x^3 = 12$.

5. Draw the graph $y = x + 2$ for the values $-3 \leq x \leq 3$.

6. Fill in this table and hence draw the graph of $y = -x$.

x	-3	-2	-1	0	1	2	3
y							

7. Complete this table and hence draw the graph of $y = 0.5x^2 + 2$.

x	-4	-3	-2	-1	0	1	2	3
$0.5x^2$								
+2								
y								

Level 7 – Algebra

> Use algebraic expressions to represent the rules for producing number patterns.

Know that most sequences are based upon addition, subtraction, multiplication or division. In more difficult cases the sequence may be based upon squares or cubes.

Example 1 Rule based on square numbers, ie the terms in the series are based on n^2.

$$2, 5, 10, 17, 26, \ldots$$

the rule is $n^2 + 1$

third term $= 3^2 + 1 = 10$, fifth term $5^2 + 1 = 26$.

Example 2 Rule based on cubed numbers, ie n^3.

$$4, 11, 30, 67, 128, 219, \ldots$$

the rule is $n^3 + 3$

third term $= 3^3 + 3 = 30$, fifth term $= 5^3 + 3 = 128$.

Example 3 More complicated rules.

$$\frac{2}{5}, \frac{3}{6}, \frac{4}{7}, \frac{5}{8}, \frac{6}{9}, \ldots$$

the rule is $\dfrac{n+1}{n+4}$.

> When you use algebraic notation you need to think in abstract terms, ie you need to use your reasoning powers to determine the solutions.
>
> Many people find it very difficult to manipulate letters (algebra). However, you should remember that the letters represent numbers, and algebra obeys all of the normal rules of number.

i. The rule for a sequence is $(n^2 + 3)$. Write the first four terms of the sequence.

   Method   when $n = 1 \rightarrow 4$

                 when $n = 2 \rightarrow 7$

                 when $n = 3 \rightarrow 12$

                 when $n = 4 \rightarrow 19$.

   Solution   *4, 7, 12, 19.*

ii. What is the general rule for this sequence?

$$4, 7, 10, 13, 16, \ldots$$

   Solution   *3n + 1.*

Level 7 – Algebra

Understand reciprocals. Know that the reciprocal of a number can be found by pressing the 1/x key on a calculator.

a. Know that a number multiplied by its reciprocal equals 1.
b. Know that $1 \div n = \frac{1}{n}$ (ie the reciprocal of n).
c. Know that we can invert a number (ie turn it upside down) to find the reciprocal.

Eg $\frac{2}{3}$ reciprocal is $\frac{3}{2}$. The reciprocal of 3 is $\frac{1}{3}$ (because 3 means $\frac{3}{1}$).

These skills are transferable. The ability to 'break' a sequence is the same ability used in general problem solving. It will help you to develop a logical mind.

What are the reciprocals of these numbers?

a. $\frac{2}{7}$ b. 81   c. 0.7.

Solution   a. $\frac{7}{2}$ (or 3.5)   b. $\frac{1}{81}$ c. $\frac{10}{7}$ (or $1\frac{3}{7}$).

Investigate complex sequences of numbers such as those generated by a computer or sophisticated calculator.

Know a variety of investigational techniques to find the rule which produces a sequence.

a. If a number sequence is alternately positive and negative, eg –3, 4, –5, 6, –7, 8, ..., there is a negative number in the rule.
b. If the numbers converge (ie get closer and closer together), eg 10, 4, 7, 5.5, 6.25, 5.875, ..., the sequence is usually based upon finding the mean (average) of the previous two numbers, ie the next number is halfway between the previous two numbers.

Breaking complicated sequences is not easy. A logical mind and perseverance are needed. The ability is only learned through practice.

Level 7 – Algebra

a. What is the rule for this sequence?
$$-4, 7, -10, 13, -16, \ldots$$
Solution   The rule is $(-1)n\,(3n+1)$.
b. What is the next term in this sequence?
$$10, 15, 12.5, 13.75, \ldots.$$
Solution   13.125.

Use the mathematical rules for simplifying mathematical expressions where the power is a positive whole number.

Know the following rules of indices (**note** integers are whole numbers).

a. Expressions can only be added or subtracted if their indices are the same (note indices are the powers, in this example the '3's).
$$4y^3 + 3y^3 = 7y^3$$
but $4y^3 + 3y^2$ cannot be simplified.

b. $4y^3 \times 3y^2 = 12y^5$.   (***Rule:*** multiply the whole numbers, add the indices.)

c. $20y^5 \div 4y^3 = 5y^2$.   (***Rule:*** divide the whole numbers, subtract the indices.)

d. $(4y^5)^3 = 64y^{15}$.   (***Rule:*** cube the whole number, multiply the indices.)

Indices are the small numbers attached to the integers. Indices are also called powers.

Simplify these algebraic expressions.

a. $5a^4 + 3a^4$ c. $8a^3 \div 4a$
b. $5c^3 \times 2c^4$ d. $(6x^3)^2$

Solutions   a. $8a^4$ c. $2a^2$ (**Note** $4a$ means $4a^1$)
               b. $10c^7$ d. $36x^6$.

Have a basic knowledge of the use and meaning of the inequality symbols shown below.

a. Know the meaning of the following symbols.
  > greater than
  < less than
  ≥ greater than or equal to
  ≤ less than or equal to

Level 7 – Algebra

b. Know that inequalities obey the same rules as equations. (The exception is that when multiplying or dividing by a negative number the inequality sign changes, ie > would become <, < would become >.)

a. Mathematics uses many shorthand symbols, eg +, −, ×, ÷, etc. Inequalities mean one term is greater or less than the other term.

b. You must fully understand equations before attempting to manipulate inequalities.

a. Place the greater than or less than sign between these two numbers so that the inequality makes sense.

$$3 \quad 8.$$

Solution $3 < 8$, ie 3 is less than 8.

b. Find the whole number values of x such that

$$15 \geq 5x > -20.$$

Method Divide the expression by 5.

This gives $3 \geq x > -4$.

Solution 3, 2, 1, 0, −1, −2, −3.

Use a trial and improvement method to solve a variety of complicated equations.

a. Know that a polynomial equation is an equation in x where at least one expression is a power of x, eg

$$x^3 - 2x = 4.$$

b. Know how to use 'trial and improvement' methods to find the solutions to a polynomial equation such as

$$x^2 - x = 7.$$

See **Algebra Level 6** for an explanation of the 'trial and improvement' method.

Use a trial and improvement method to find the solution to this equation.

$$x^2 - x = 7.$$

Solution $x = 3.193$ (approx).

Use algebraic elimination or substitution methods to solve a pair of simultaneous linear equations with two unknowns.

Level 7 – Algebra

Know how to solve two simultaneous equations such as

$$4x - y = 18$$
$$\text{and } 3x + 2y = 19$$

Method: there are two variables, x and y. One variable must be eliminated. Multiply the first equation by 3, the second equation by 4. Then subtract equation two from equation one. The x terms disappear leaving a single linear equation to solve. (See below.)

> Simultaneous equations are connected. The two equations have identical values for the x, and also for the y, which work with both equations. Simultaneous equations can also be solved by graphical and matrix methods.

Solve $4x - y = 18$, and $3x + 2y = 19$.

Method $4x - y = 18$ (multiply by 3) $12x - 3y = 54$
$3x + 2y = 19$ (multiply by 4) $12x + 8y = 76$
Subtract $-11y = -22$
$y = 2$

Substitute in first equation (ie $4x - y = 18$)

this gives $x = 5$.

Solution $y = 2, x = 5.$

> Draw and use a variety of graphs, eg conversion graphs.

a. Be able to draw and use a conversion graph, eg US dollars to UK pounds.

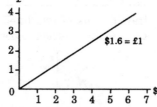

b. Be able to solve a distance/time problem by graphical methods, eg this graph shows the distance a car travels.

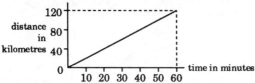

c. Be able to draw a graph to show information.

Level 7 – Algebra

> It is often easier to solve mathematical problems using graphical methods rather than calculation methods. The answers are, of course, the same.

a. Given that 0°C = 32°F, and 20°C = 68°F, draw a graph to convert °C to °F. Use this graph to convert 15°C into °F.

Solution   *59°F.*

b. A car travels at a speed of 120 km/h. Draw a distance/ time graph to show this and from the graph find the time the car takes to travel 80 kilometres.

Solution   *The graph should be the same as the one in the box above. Answer 40 minutes.*

c. A car starts from A, it travels at a speed of 60 km/h for 20 minutes, stops for 10 minutes and then continues for 15 minutes at a speed of 40 km/h. Draw a graph to show this.

Solution

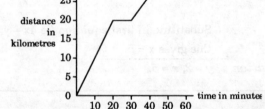

> Solve a pair of simultaneous linear equations with two unknowns by a graphical method.

Be able to draw graphs of two linear equations, where the dotted lines cross is the solution to both equations.

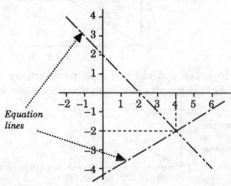

Solution $x = 4, y = -2$

Level 7 – Algebra

This requires a graphical solution to simultaneous equations. A calculator method was shown earlier in this section. If the question specifies a method by which a question must be solved, eg 'solve by graphical methods', then that method must be used.

Draw the graphs of the simultaneous equations
$$2x + y = 5$$
$$\text{and } 3x + 2y = 8$$
and hence find the values of x and y which satisfy both equations.

Solution $x = 2, y = 1$

Use computers or sophisticated calculators to draw graphs.

Be able to use a computer to draw a graph.

Computers are being increasingly used in mathematics.

Draw the graph of $y = 4x^3 - 3$ using a computer.

59

Level 7 – Algebra

Attainment Test 7/3   *Answers on page 153*

1. The rule to produce a sequence of terms is given as $(n^3 - 1)$. Write down the first 4 terms of this sequence.
2. Use symbols to express the rules for these sequences.
   (a) 3, 5, 7, 9, 11, ...
   (b) 7, 10, 15, 22, 31, ...
   (c) $\frac{3}{4}$, $\frac{4}{5}$, $\frac{5}{6}$, $\frac{6}{7}$, $\frac{7}{8}$, ...
3. What is the value of y if $y \times 5 = 1$?
4. What is the reciprocal of
   (a) 7
   (b) $\frac{1}{8}$
   (c) $\frac{3}{4}$
5. The rule for producing a sequence is $(-1)^n (n + 5)$. Write down the sixth and seventh terms in this sequence.
6. Simplify
   (a) $a^3 \times a^4$
   (b) $2a^5 \times 4a^3$
   (c) $30y^3 \div 6y$
   (d) $(3y^2)^3$.
7. Find the whole number values of y such that $-3 \leq y < 2$.
8. Use trial and improvement methods to find a solution to
$$x^2 + x = 8,$$
correct to 3 significant figures.

Level 7 – Algebra

Attainment Test 7/4 *Answers on page 154*

1. Solve these simultaneous equations. *(Do not use graphical methods.)*
$$3a + 4y = 15$$
$$2a - 2y = -4$$

2. Solve these simultaneous equations. *(Do not use graphical methods.)*
$$5c = 3a + 1 \text{ and } 4c + 2a = 14$$

3. 4.5 litres = 1 gallon. Using this information draw a conversion graph for gallons to litres. Use your graph to answer the following questions.

    (a) How many litres are equivalent to 3 gallons?

    (b) How many gallons are equivalent to 14 litres?

    (c) My car travels 90 miles on two gallons of petrol. How far would it travel on two litres of petrol?

4. Use graphical methods to solve these simultaneous equations.
$$3x + y = 10$$
$$x + 2y = 5$$

5. The length of a square room was given as 4.7 m, correct to the nearest 0.1 m. What are the maximum and minimum limits of the area of the room?

Shape & Space

Shape and space are both important features of the world in which we live. You need to be able to describe and distinguish a variety of shapes. You should also be able to locate objects, whether it be by simple use of words such as up, down, to the left, etc, or using more complex mathematical directions such as co-ordinates (x,y,z) to describe the location of a point in three-dimensional space.

Key

 The blackboard represents the level of attainment required.

 The dog holding a calculator shows the skills and knowledge required for this topic.

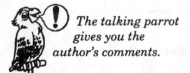 *The talking parrot gives you the author's comments.*

The question mark stands for typical tests and questions that you will have to complete.

Level 4 – Shape and space

Have a basic understanding of the sizes of angles and be able to draw and measure angles up to 360°.

a. Know the following mathematical terms:

Acute angle 0° to 90°
Obtuse angle 90° to 180°
Reflex angle 180° to 360°
Right angle equals 90°
Angles on a straight line add up to 180°
Angles at the centre of a circle add up to 360°.

b. Know the following types of line:

parallel lines (same distance apart at all points)
perpendicular lines (at right angles to each other)
vertical lines
horizontal lines.

a. You need to be familiar with the different types of angle. You should be able to make a reasonable estimate of the size of any angle.
b. You should be familiar with these mathematical terms.

a. What is the size of angle x?

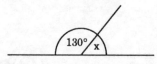

Solution   50°.

b. Look at this shape and answer the following questions.

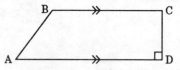

i. Which lines are parallel?
ii. Which line is perpendicular to AD?
iii. Name a line which is horizontal.
iv. Name a line which is vertical.

Solution   i. AD and BC   ii. CD
             iii. AD (or BC)   iv. CD

Level 4 – Shape and space

> Be able to draw two-dimensional (2-D) shapes and able to make three-dimensional (3-D) shapes.

a. Know how to use a pair of compasses, ruler and pencil to construct the following – equilateral triangle, isosceles triangle, bisect an angle, bisect a line, construct a perpendicular at a given point, rectangle, square, circle, cube, cuboid, prism, cylinder, square-based pyramid, triangular-based pyramid.

b. Know how to recognise a shape from its net or two-dimensional plan and be able to draw the nets of the following – cube, cuboid, prism, cylinder, square-based pyramid, triangular-based pyramid.

> a. You need to be able to produce accurate and neat constructions.
>
> b. You need to acquire the ability to recognise common 3–D shapes from their 2–D nets.

a. Construct a triangle with the sides 4cm, 5cm and 6cm.
b. i. Name the 3–D shape this net would make.

Solution   *Cylinder.*

ii. Draw an accurate net of a cube, side 3 cm.

> Use co-ordinates (x,y) to specify the position of a point (x and y will both be positive.) Specify the position of a point by giving its distance and direction from another point. Understand compass directions North, South, East and West.

a. Be able to locate places on a map or in an atlas by using map references.
b. Be able to locate a place on a map using bearings and distance.
c. (If your school uses LOGO computer commands) know how to use the commands for distance and direction.

Level 4 – Shape and space

a. On Ordnance Survey maps, grid references are often given as a 4- or 6-figure number. On street maps, the map is often divided into squares, eg A3. You need to be able to use various forms of grid reference.

b. In navigation, bearings and distances are used to determine position. Bearings are also needed when walking and using a compass.

c. The ability to use a computer is an advantage in the modern world.

a. What is the name of a certain village at map reference ...? (A 4- or 6-figure grid reference would then be given.)

b. Axeford is 80 km from Bedworth on a bearing of 070°. Mark the position of Axeford.

Method: Use a protractor to find the direction, note the scale of the map and mark Axeford.

c. You are expected to use LOGO commands for distance and direction.

GET SMART

Recognise two dimensional shapes which have a symmetry when they are rotated.

a. Understand the term rotation.

b. Know how to test for rotational symmetry and understand the terms:

rotational symmetry of order 2
rotational symmetry of order 3
rotational symmetry of order 4, etc

(*Note:* If a shape is turned through 360° and fits exactly on top of itself more than once, it has rotational symmetry, and the number of times it fits exactly on top of itself is its order of symmetry.)

a. Rotation is the turning of a shape. One complete rotation means a turn through 360°.

b. It is advisable to use tracing paper. Trace the shape, turn the tracing paper through 360° and count the number of times the shape fits exactly on top of itself. Eg a square fits exactly on top of itself four times, therefore it has a rotational symmetry of order 4.

a. Which type of triangle has rotational symmetry?

Solution   Equilateral triangle (rotational symmetry of order 3).

Level 4 – Shape and space

b. Does this shape have rotational symmetry, and if so of what order?

Solution   The shape has rotational symmetry of order 2, because in turning the shape through 360° it fits on top of itself twice.

Estimate areas by counting squares. Estimate volume by counting cubes.

Know how to find the approximate area and volume of an irregular shape.

You should draw a square grid and then find the area by counting squares.

Draw a centimetre square grid and use this to find the approximate area of your hand.

Solution   An Answer between 70 cm² and 150 cm².

Be able to find the perimeter of simple shapes.

You should know that perimeter means the distance around a shape and be able to calculate that distance.

If we need to calculate the amount of wall paper for a room we need to find the perimeter as part of our calculation.

Find the perimeter of this rectangle.

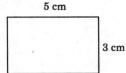

5 cm

3 cm

Method   Perimeter = 5 + 3 + 5 + 3
Solution   16 cm.

67

Level 4 – Shape and space

Attainment Test 4/5

Answers on page 143

1. In this shape which line is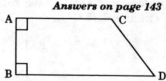
   (a) parallel to BC?
   (b) perpendicular to BC?

2. What are the special names given to these angles?

   (a)   (b)   (c)

   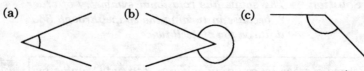

3. AB is a straight line.
   What is the value of x?

   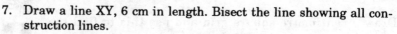

4. What is the size of the angle between two perpendicular lines?

5. Using pencil, ruler and a pair of compasses construct an equilateral triangle with sides 6 cm.

6. Copy this drawing. Bisect the angle ABC showing all construction lines.

   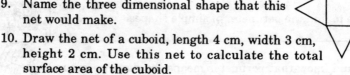

7. Draw a line XY, 6 cm in length. Bisect the line showing all construction lines.

8. Draw a triangle ABC with sides as follows:
   AB = 4 cm, AC = 5 cm, BC = 3 cm.
   Measure angle B.

9. Name the three dimensional shape that this net would make.

10. Draw the net of a cuboid, length 4 cm, width 3 cm, height 2 cm. Use this net to calculate the total surface area of the cuboid.

11. Write down the order of rotational symmetry of the following shapes:
    (a) Rectangle (b) Square (c) Regular Hexagon (d) Circle.

12. Appleton is 40 km from Barbridge on a bearing of 080°. The scale of the map is 1 cm represents 10 km. Mark the position of Appleton.

    • Barbridge

13. The perimeter of a rectangle is 26 m. The length of the rectangle is 8 m. What is the length of the other side?

Level 5 – Shape and space

> Recognise shapes which are identical to each other, ie shapes which are congruent.

Understand the meaning of congruency.

> Congruent shapes are shapes which are identical. If you place one shape on top of the other the shapes will fit exactly.

Which of there shapes are congruent?

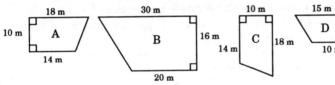

Solution A is congruent to C (because the sides and angles of A are exactly the same size as corresponding sides and angles of C).

> Understand the angle properties of straight lines, triangles and parallel lines and use these properties to calculate the size of angles.

a. Know angles on a straight line add up to 180° and vertically opposite angles are equal.

   Example $a + b = 180°$ (straight line)
   $b + c = 180°$ (straight line)
   $a = c$ (vertically opposite)

b. Know angle properties of parallel lines.
   In the diagram:

   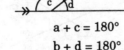

   $a = d$ $a + c = 180°$
   $b = c$ $b + d = 180°$

c. Know how to use the properties of parallel lines when solving parallelogram and trapezium questions.

d. Know the angle properties of triangles.

   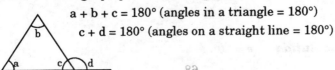

   $a + b + c = 180°$ (angles in a triangle = 180°)
   $c + d = 180°$ (angles on a straight line = 180°)

Level 5 – Shape and space

Equilateral triangle
All sides equal; each angle equals 60°.

Isosceles triangle
Two sides equal; two angles equal.

a. Parallel lines occur frequently in examination questions.

b. You should be able to use the above properties. They are frequently encountered as small parts of more difficult questions.

c. You must know these angle properties. They are frequently encountered as small parts of more difficult questions.

a. i. What is the value of a?

Solution 60°.

ii. What are the values of a, b and c?

Solutions $a = 130°$ (angles on a straight line)
$b = 50°$ (vertically opposite)
$c = 130°$ (vertically opposite to a).

b. i. What is the value of x?

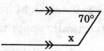

ii. What is the value of y?

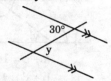

Solutions   i. $x = 110°$ ii. $y = 30°$.

c. What is the value of a?

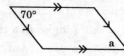

Solution $a = 70°$.

Level 5 – Shape and space

d. In each question find the value of x.

i.

ii.

iii.

iv.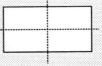

Solutions i. 80° ii. 70° iii. 60° iv 20°.

Understand that certain shapes are symmetrical, ie a line can be drawn on the shape such that a mirror image appears. For example, this rectangle has two axes of symmetry, indicated by the dotted lines. Place a mirror on the dotted lines to see the symmetry. You should know where to draw the axes of symmetry in common shapes.

Know the symmetrical properties of the following shapes:

Shape	Axes of symmetry	Shape	Axes of symmetry
Isoceles triangle	1	Rhombus	2
Equilateral triangle	3	Parallelogram	0
Square	4	Kite	1
Rectangle	2		

The centre of symmetry is the place where two or more lines of symmetry cross (ie in the middle of the shape).

How many axes of symmetry has a 50p coin?
Solution 7.

You should be able to use arrow diagrams (called networks) to solve problems and to present information.

Know how to draw a network to simplify a problem. Questions are often more easily understood as a network, eg this network shows the bus routes between four towns.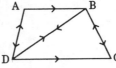

Level 5 – Shape and space

You must be able to present and understand information in a variety of forms.

Look at the network in the box above. How would a person get from C to A by bus.

Solution *The only way is from C to B to D to A.*

*(**Note:** If distances had been given on the network diagram, as they often are, then the distance from C to A by bus could have been calculated.)*

Use a protractor to draw and measure angles of various sizes.

a. Use a protractor to measure an angle.
b. Use a protractor, ruler and pencil to draw an angle.
c. An acute angle is between 0° and 90°;
   an obtuse angle is between 90° and 180°;
   a reflex angle is between 180° and 360°.

a. This is a test of accuracy (to the nearest degree). It also tests your ability to read scales.

b. This demonstrates the ability to use geometrical instruments. Neatness is essential. Untidy or messy drawings lose marks. This is a basic skill which will be needed later when more complex constructions are undertaken.

c. The protractor has two scales. A knowledge of acute, obtuse and reflex angles will help you to determine which scale to use in different circumstances.

a. Use a protractor to measure this angle.

Solution 161°.

b. Draw an angle of 72°.

Solution

Level 5 — Shape and space

c. Estimate the size of these angles to the nearest 10° without using a protractor.

(a)   (b)

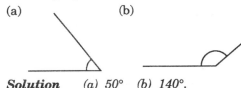

Solution (a) *50°* (b) *140°*.

Be able to find the area of simple 2-D shapes.

You are expected to be able to use these formulae.

Area of a rectangle = Length × Width

Area of a triangle = $\frac{1}{2}$ Base × Perpendicular height

You have to calculate the area of a room if you want to lay carpet. (*Note:* Area is measured in squared units, eg m², cm², mm², etc.)

i. What is the area of a square, side 8 cm?

Solution $8 \times 8 = \textbf{64 cm}^2$,

ii. What area of carpet is required for a room 6 m long and 5 m wide?

Solution $6 \times 5 = \textbf{30 m}^2$.

iii. What is the area of this triangle?

Solution $\frac{1}{2} \times 12 \times 7 = \textbf{42 cm}^2$.

Be able to use formulae to find the volume of simple solids.

You are expected to be able to use these formulae.

Volume of a cuboid (or cube) = Length × Width × Height

Volume of a prism = Area of cross-section × length

When packaging goods, we often need to find the volume of a box to calculate how much will fit inside. (*Note:* A cube is a cuboid with sides of equal length.)

Level 5 – Shape and space

  i. Find the volume of a cube, side 3 cm.
    Solution $3 \times 3 \times 3 = 27$ cm².
  ii. Find the volume of this prism.

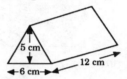

Method   The area of cross-section (ie the triangle) is
                    $\frac{1}{2} \times$ base $\times$ height
                    = $\frac{1}{2} \times 6 \times 5 = 15$ cm²

Volume   = area of cross-section $\times$ length
                    = $15 \times 12 = 180$ cm³

Solution   180 cm³.

Have a basic understanding of π and the way in which it connects the diameter and circumference of a circle.

Be able to find a connection between the diameter of a circle and its circumference. The connections are

i. circumference ÷ diameter = π
ii. diameter × π = circumference.

Note: the value of π is 3.142, correct to 3 decimal places (or approximately 3).

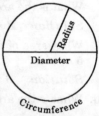

π is one of the most widely used mathematical terms. It is important to understand that π is the connection between the length of the diameter and the length of the circumference of a circle. For most practical purposes we can assume that the length of the circumference is approximately three times the length of the diameter.

A circle has a diameter of 4 m. What is the approximate length of the circumference?

Method   We know that the diameter $\times \pi$ = circumference.
                π is approximately 3.
              $4 \times 3 = 12$.

Solution   12 m (approx).

Level 5 – Shape and space

Attainment Test 5/6   *Answers on page 147*

1. Which of these triangles are similar?

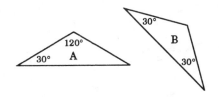

2. Which of these triangles are congruent?

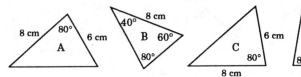

3. AB is a straight line. What is the value of x?

4. Find the values of a, b and c.

5. Find the value of x.

6. What is the value of y?

7. Find x.

Level 5 – Shape and space

8. What mathematical statement can be made about the lines AB and CD?

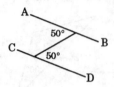

9. Find the values of x.

   (a)   (b)   (c)

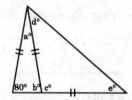

10. What are the values of a, b, c, d and e?

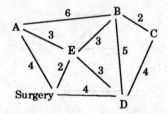

11. How may axes of symmetry do the following shapes have?

    (a) Square (b) Rhombus (c) Parallelogram (d) Circle.

12. A doctor has to visit several patients. The distances (in kilometres) between the houses are given in the diagram.

    The doctor must start and finish at the surgery. What route should he take to minimise the distance he has to travel and what is this distance?

Level 5 – Shape and space

Attainment Test 5/7   *Answers on page 148*

Questions 1–4. Use your protractor to measure these angles.

1.   2.

3.   4.

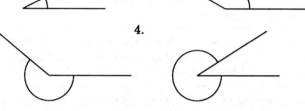

5. Use your protractor to measure each of the angles in this triangle.

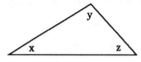

6. (a) Draw a triangle ABC such that angle A = 30° and angle B = 70°. The line AB is drawn.

   A ———————————————— B

   (b) What is the size of angle C?
   (c) If an ant crawled round the triangle how far would it crawl?

7. The area of a square is equal to the area of this triangle.

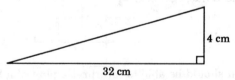

   What is the perimeter of the square?

8. The volume of this prism is 320 cm². The cross-sectional area is 64 cm².

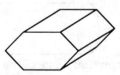

   What is the length of the prism?

9. What is the approximate diameter of a circle with a circumference of 60 cm?

Level 6 – Shape and space

Make use of the angle and symmetry properties of 4 sided shapes and other shapes (often called polygons) to solve mathematical problems.

Know that if shapes are to tessellate (ie fit together) the sum of their angles at a point must add up to 360°.

A polygon is a shape. A regular polygon is a shape in which all of the sides and all of the angles are equal, eg a square could be described as a regular 4-sided polygon.

How many squares and how many regular octagons (8-sided) are required to tessellate at a point?

Solution *One square and two octagons. (Each angle in an octagon is 135°. 90° + 135° + 135° = 360°.)*

Able to draw three dimensional shapes on paper.

a. Be able to draw 3-D shapes, eg a cuboid.
b. Be able to draw nets and recognise nets of common 3-D shapes.
c. Understand simple plans and elevations.

a. This ability shows an understanding of shape and space.
b. A net is the 2-D representation of a 3-D shape when placed flat, eg a box.

c. You should be able to interpret a plan of a home extension.

a. Draw a sketch of a cylinder.
Solution

b. Give the mathematical name of the shape formed by this net.

Solution *Triangular-based pyramid.*

c. Calculate the volume and surface area of a shape from its plan.

Level 6 – Shape and space

> Draw a variety of shapes using computers.

Be able to use LOGO to draw shapes.

> Computers can be more efficient for drawing shapes than humans using conventional drawing instruments.

Use LOGO to draw a regular octagon.

> Know the special mathematical names and properties of four-sided shapes (called quadrilaterals).

Know the properties of these quadrilaterals.

Square (4 equal sides, 4 equal angles, opposite sides parallel)

Rectangle (4 equal angles, opposite sides equal and parallel)

Parallelogram (opposite sides are equal and parallel, opposite angles are equal, diagonals bisect each other)

Rhombus (a parallelogram with all four sides equal length)

Trapezium (one pair of parallel sides)

Kite (two pairs of equal sides)

> You need to recognise common shapes. It is important to be able to distinguish between the different quadrilaterals shown above.

What is the special name of this shape?

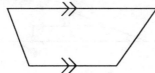

Solution *Trapezium.*

Level 6 – Shape and space

 Use bearings from 000° to 360° to indicate direction.

Be able to calculate the bearing from one place to another, and plot the course of a ship or plane. This involves using bearing together with scale to produce a scale drawing of the flight of an aircraft.

 You should know how to use a ruler and protractor to measure bearings. All bearings are calculated by measuring from due North and all bearings are written as a three figure number, eg 90° is written 090°. Bearings can be between 000° and 360°.

What is the bearing of A from B?

Solution   120°.

 Understand reflection, mirror image and symmetry.

Be able to select the appropriate position for a mirror on a shape so as to create a mirror image.

 You should be able to draw a line of symmetry on a symmetrical shape.

AB is a line of symmetry. Complete the drawing.

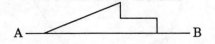

Solution   *Place a mirror along AB to check.*

Level 6 – Shape and space

| Be able to increase the length of a shape by a scale factor, eg make all sides of a shape three times larger. |

a. Be able to enlarge a shape by a scale factor, eg X is an enlargement of Y, scale factor 2.

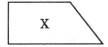

b. Be able to enlarge a shape with a specified centre of enlargement, eg T is an enlargement of S, scale factor 2, centre of enlargement the origin, ie (0,0).

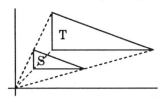

a. *Note*, X is the same shape as Y, all angles are equal, sides are twice the length of Y.

b. The method for enlargement is first to draw dotted lines from the centre of enlargement to each corner of the shape. Then, for a scale factor of 2, double the length of the dotted lines to produce T.

Enlarge a 2 cm square by a scale factor of 4.
a. What is the length of each side?
b. What is the area?
Solution   *a. 8 cm.   b. 64 cm².*

| Be able to use a computer to draw lines and shapes. |

a. Be able to give clear instructions which a computer can understand.

b. Be able to select a suitable instruction from a group of instructions provided by the computer to achieve the desired result.

| The ability to use a computer for complex tasks is becoming increasingly important in the modern world. The computer can save a considerable amount of time. |

Level 6 – Shape and space

Be able to use the area and circumference formulae for circles.

a. Be able to find the area and circumference of a circle given the radius (or diameter). The radius is half the diameter.
b. Be able to find the radius or diameter of a circle given the area or circumference.

This is a practical application of using formulae of the type met in the ***Algebra*** section. You are expected to be able to use formulae and have a good understanding of circles. You must memorise these formulae.

Area = πr^2 Circumference = $2\pi r$ (r = radius,
                                   = πd d = diameter)

a. The area of a circular pond is 23 000 m². A man rows across the pond from one side to the other (ie the diameter). How far does the man row?

Method Area = πr^2

$$r^2 = \frac{\text{Area}}{\pi}$$

$$r = \sqrt{\frac{\text{Area}}{\pi}}$$

$$r = \sqrt{\frac{23\ 000}{3.142}} = 85 \text{ m or } 86 \text{ m (approx)}$$

The radius is half the diameter.
Solution *The diameter is about 170 m.*

b. The diameter of a circular pond is 50 m. How long will it take a man to run around the edge of the pond twelve times at a speed of 1.8 m per second?

Method   Circumference = $2\pi r = 2 \times \pi \times r = 2 \times \pi \times 25$
                            = 157 m (approx).
          12 laps means 157 × 12
                            = 1 885 m (total distance).
          1 885 ÷ 1.8   = 1 047 seconds
Solution *17 minutes 27 seconds (approx).*

Level 6 – Shape and space

Attainment Test 6/4 *Answers on page 151*

1. John wishes to use squares and equilateral triangles for a tessellation pattern. How many squares and how many triangles must he use to tessellate at a point?

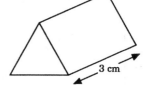

2. Draw the net of this shape. The ends are equilateral triangles, side 2 cm.

3. Which three-dimensional shape can be produced from this net?

4. Give the mathematical names of these shapes.

   (a) *All four sides are equal lengths.* (b) (c)

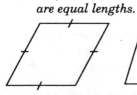 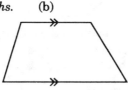 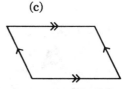

5. A ship starts from A. It sails due North for 20 km to B. It then alters course to a bearing of 075° and sails for 40 km to C. It then changes direction to a bearing of 190° and sails for 30 km to D.

   (a) Using a scale of 1 cm represents 5 km draw a scale diagram to show the course of the ship.
   (b) What is the distance from A to D? (Give your answer in kilometres.)
   (c) What is the bearing of point A from point D?
   (d) What is the bearing of point D from point A?

6. The diagram at the top of the next page shows the position of Arden and Blandon. The scale of the map is 1 cm represents 20 km.

   (a) What is the bearing of Arden from Blandon?
   (b) What is the distance from Blandon to Arden in km?
   (c) Carlton is 60 km from Arden and 140 km from Blandon. Copy the diagram and mark the position of Carlton.
   (d) What is the bearing of Carlton from Arden?

Level 6 – Shape and space

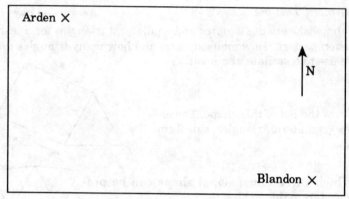

7. The line XY is a line of symmetry. Complete the shape.

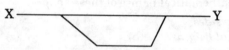

8. How many lines of symmetry do the following letters have?

   A   C   H   Z

9. What is the scale factor of the enlargement required to transform Shape A into Shape B?

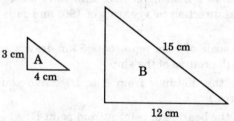

10. Rectangle A has co-ordinates (4,2), (6,2), (4,3) and (6,3). Draw this rectangle on squared paper and enlarge it by a scale factor of 2, centre of enlargement the point (2,1). Give the co-ordinates of the rectangle which is formed by this enlargement.

11. Enlarge this triangle by a scale factor of 2. Centre of enlargement is the point A. What is the area of the enlarged triangle?

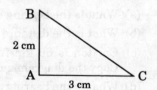

12. The circumference of a circle is 50.24 m. Using π = 3.14 calculate the area of the circle.

Level 7 – Shape and space

> Use Pythagoras' Theorem to find the length of the third side of a right angled triangle when given the lengths of the other two sides.

Be able to calculate the length of the third side of a right-angled triangle given two sides.

Memorise $a^2 = b^2 + c^2$

Method

To find the **Long Side**
Square the two sides
Add
Find the square root

To find a **Short Side**
Square the two sides
Subtract
Find the square root

> Pythagoras' Theorem is the introduction to trigonometry. It is vital to understand this section if you are to proceed to Level 8.

i. Find x.

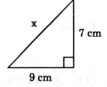

Method $9^2 + 7^2 = 81 + 49 = 130$.
Square root of $130 = 11.4$ cm

Solution   11.4 cm.

ii. Find x.

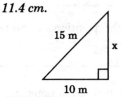

Solution   11.18 m.

> Able to draw the locus (or path) taken by an object when moving according to given rules or instructions.

Know that when a point moves in such a way that it obeys certain conditions the path which that point takes is called a locus (plural loci), eg this is the locus of a point which is always 1 cm from the line AB.

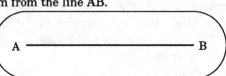

Level 7 – Shape and space

To answer loci questions you need an elementary understanding of geometry. You should know how to bisect an angle and a line.

Draw the locus of a point which is always 3 cm from the point A. What shape is formed?
Solution *A circle radius 3 cm about the point A.*

Enlarge a shape by a fractional scale factor.

Know how to enlarge a shape by a fractional scale factor. (Look at the second diagram on page 79, ***Shape and Space, Level 6***. If we enlarge triangle T by a scale factor of $\frac{1}{2}$ it will become triangle S. The method is the same, except the length of the dotted line is halved.)

When enlarging by a fractional scale factor the shape will become smaller.

Enlarge a square, side 12 cm, by a scale factor of $\frac{1}{3}$.
Solution *A square, side 4 cm.*

Describe and identify the position of points in 3-D using (x,y,z) co-ordinates.

Be able to think of solids in three dimensions.

You already know how to use co-ordinates (x,y) to locate points in two dimensions. By the addition of z, you can locate an object in space, ie three dimensions, because the object has length, width and height.

Given that the following co-ordinates are the vertices (corners) of a cuboid, find its dimensions and hence its volume.
(0,0,0), (3,0,0), (3,0,2), (0,0,2), (0,4,0), (3,4,0), (3,4,2), (0,4,2).
Solution *The dimensions are 3 by 4 by 2; the volume is 24 cubic units.*

Level 7 – Shape and space

Be able to find the area and volume of a variety of 2-D and 3-D shapes.

Know how to find the area and volume of the following shapes and be able to combine the shapes to find the total area and volume. Shapes needed are:

2-D rectangle, square, triangle, parallelogram, circle
3-D cube, cuboid, cylinder, prism, solids of uniform cross-sectional area.

You already know how to find the areas and volumes given above. You now need to combine these shapes to produce the areas and volumes of complex shapes such as you would meet in real life situations.

a. What is the area of this running track? The two ends are semi-circles.

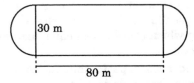

Method Two semi-circles make one full circle. The radius of the circle is 15 m.
Area of rectangle = 30 m × 80 m = 2 400 m²
Area of circle = πr^2 = π × 15 × 15 = 707 m²
Total area = 2 400 m² + 707 m² = 3 107 m²

Solution 3 107 m².

b. Find the volume of this shape. (*Note* the shaded area is the cross-section.)

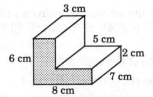

Method Volume = Area of cross-section × length
= 28 cm² × 7 cm = 196 cm³

Solution 196 cm³.

Level 7 – Shape and space

Attainment Test 7/5
Answers on page 154

1. Find the length of AB.

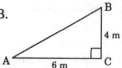

2. Find the length of AC.

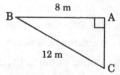

3. ABCD is a rectangle. Find the length of AC.

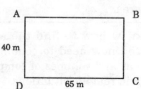

4. Draw the locus of a point which is 4 cm away from the outside edge of a circle of radius 2 cm.

5. Two lines, AB and AC, meet at the point A. The angle BAC is 40°. The length of each line is 5 cm. Draw the two lines and draw the locus of a point which is always an equal distance away from the lines AB and AC.

6. A goat is tethered by a rope to the L-shaped rail shown. The rope is 3 m long. Using a scale of 1 cm represents 1 m, draw an accurate scale diagram and shade the area within which the goat can move.

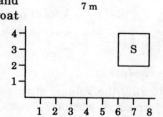

7. S' is the image of S after an enlargement, scale factor 0.5. The centre of enlargement is the origin. Copy the diagram and draw the position of S'.

8. These are the three dimensional co-ordinates of a box. A (0,0,0), B (3,0,0), C (3,0,4), D (0,0,4), E (0,2,0), F (3,2,0), G (3,2,4), H (0,2,4).

    (a) What is the volume of the box?

    (b) What is the total surface area of the box?

    (c) Give the co-ordinates of the point where the lines AG and BH intersect.

9. The volume of a cube is 729 cm³. A circular hole is cut out of the cube such that the diameter of the hole is x cm and the depth of the hole is 2x cm. Given that the length of each side of the cube is 3x cm, calculate the volume of the circular hole.

Handling Data

Mathematics provides a powerful means of communication. Every day you make use of mathematical data, probably without realising it, eg bus and train timetables, football league tables, etc. Mathematical information can be presented in a variety of forms such as graphs. diagrams, flow charts, tables, etc. It is important to be able to interpret information, irrespective of the form which it takes. It is also essential to be able to communicate information using graphs, diagrams, etc.

Key

 The blackboard represents the level of attainment required.

 The dog holding a calculator shows the skills and knowledge required for this topic.

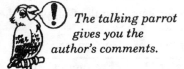 *The talking parrot gives you the author's comments.*

? *The question mark stands for typical tests and questions that you will have to complete.*

Level 4 – Handling data

Select a problem situation which can be solved by a mathematical survey (eg the shoe sizes of the pupils in a class) collect the data (ie information) and record it in a table, using tallying methods (test results such as those at the bottom of this page).

a. Be able to carry out a simple statistical survey and select a suitable collection method, eg questionnaire, show of hand, extraction of data from charts or tables.

b. Be able to order data. If test results are obtained at random some form of ordering is required. You should be able to place data in order, eg smallest to largest.

c. Be able to collect data using a tally system and produce a frequency table with, eg, test marks 1 – 10, 11 – 20, 21 – 30, 31 – 40, etc.

a. In the world of work, people are often asked to produce, make use of, or analyse data. The technique is only learned by practice in real situations. You will be asked to undertake projects or surveys, eg carry out a traffic survey.

b. It is far easier to understand information that is presented in an ordered form.

c. A frequency table is a popular method of presenting data.

a. You will be required to carry out a variety of surveys, eg find and record the favourite school subject of other pupils in your class. You will be required to collect data and interpret your findings.

b. These are the heights of five pupils. Present the data in an ordered form.

   1.58 m, 1.39 m, 1.72 m, 1.63 m, 1.49 m.

Solution   *1.39 m, 1.49 m, 1.58 m, 1.63 m, 1.72 m.*

c. These are the test results for some pupils. Complete the table below to show the results.

   6, 32, 28, 11, 23, 17, 24, 19, 18, 27, 29, 27

Mark	Tally	Frequency
0–10	/	1
11–20		
21–30	||||| /	6
31–40		

Solution   *Frequency 11–20 = 4, 31–40 = 1.*

Level 4 – Handling data

> Able to calculate the mean and range when given a collection of information.

Be able to calculate the mean (often called average or mean average) and range (difference between the largest and smallest value) of a set of data and be able to use the mean and range to make comparisons between data.

> This section covers the analysis of data. This is an important skill in the modern world. People frequently have to make decisions based upon a collection of data, eg buying a washing machine – the buyer must compare the qualities of a variety of machines.

Find the mean range of these two sets of data and make comments based upon your findings.

Batsman	Scores
Jones	36 42 38 57 34 63 54 57 43 48
Smith	8 103 27 92 0 38 87 85 10 25 54 41

Solution   Jones mean score $\dfrac{472}{10} = 47.2$

Smith mean score $\dfrac{570}{12} = 47.5$

Jones range $(34 - 63) = 29$

Smith range $(0 - 103) = 103$

Comment: Although Smith has a higher mean (average) score, Jones is more consistent. *(Note: The batsmen have played different numbers of innings. Therefore it would be pointless to compare total scores but we can compare average scores and the range of scores.)*

> Use a computer data base.

Be able to extract information from a computer.

> Many shops and offices are now computerised, eg travel agents check flight availability by using computers. You will learn these skills at school (in other lessons besides mathematics). In the modern technological age the ability to use a computer is becoming increasingly important. Many people now use a home computer to help them organise their finances. This involves interrogating data.

Level 4 – Handling data

Sort information into specific categories.

Be able to design a tree-diagram to sort information.

This is needed when designing computer programs. This is a decision box.

It can only have a Yes or No answer.

There are some French and English people on a ferry. Design and draw a decision-tree diagram to sort them into groups.

Solution

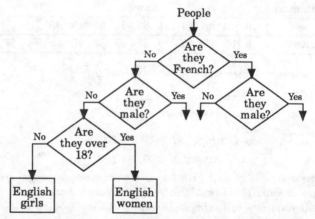

The whole decision tree diagram would be able to sort the people into the following categories: English Girls, English Women, English Boys, English Men, French Girls, French Women, French Boys, French Men.

Draw, understand and be able to extract information from a bar-line graph for discrete variables.

Be able to draw a bar line graph.

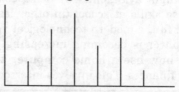

92

Level 4 – Handling data

You need to be able to present information in a variety of ways. You need to select the most appropriate graph to show the information. Statistics are usually easier to understand in a graph rather than a table. A discrete variable is a measure which has an exact value. Eg the number of pupils in each class must be a whole number: we cannot have 3.271 pupils.

But the weight of pupils can be any number, so weight is continuous not discreet.

You will be asked to draw and extract information from bar graphs.

Draw, understand and be able to extract information from line graphs.

Be able to draw a line graph.

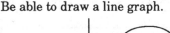

If the line graph above showed the number of pupils in a school at various times, certain values would have no meaning, ie there would never be 3.271 pupils in the school, but if the graph showed the temperature then it could be 3.271° at some time during the day.

You will be asked to draw and read information given in a line graph.

Draw, understand and be able to extract information from a frequency graph.

Be able to draw a frequency graph.

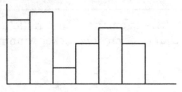

Level 4 – Handling data

You need to be able to extract information from a variety of graphs and be able to understand and make use of the information you obtain.

You will be asked to read graphs and comment upon your findings.

Have a basic understanding of probability.

a. Understand that events have different probabilities from 0 to 1.

    A probability of 0 means the event cannot happen.

    A probability of 1 means the event will happen.

b. Make a reasonable estimation of the probability of an event, choose from no chance, poor chance, even chance, good chance, certain.

a. People need to assess probability in everyday life, eg is it worth buying an old car if it will probably break down?

b. Insurance companies use probability to determine their premiums. The less likely an event the lower the premium. (*Note* the 'premium' is the annual fee you have to pay the insurance company.)

a. Give an event which has a probability of: i. 0; ii. 1.

    Solution   i. Anything which is impossible.

                    ii. Anything which is certain.

b. State the probability of a die landing on: i. 6; ii. 7; iii. an odd number. Choose from no chance, poor chance, even chance, good chance, certain.

    Solution   i. *poor chance*; ii. *no chance*; iii. *even chance*.

Estimate the probability of an event occurring and give reasons for the estimation if it is a subjective estimate.

a. Know that if there is one prize in a raffle and fifty tickets are sold the chance of any one ticket winning is 1 in 50 or $\frac{1}{50}$.

Level 4 – Handling data

b. Be able to make a subjective statement about the probability of an event happening, eg the chance of the school bus arriving late tomorrow is about 1 in 5. And be able to justify the statement, eg it is $\frac{1}{5}$ because the bus has arrived late about one day in five for the last year.

a. *Note* this is not a subjective estimate. It is based on mathematical fact. If there are 50 tickets the probability is 1 in 50.

b. *Note* this is a subjective estimate. There are numerous variables affecting the probability of the bus being late. Subjective estimate is a way of saying an 'educated guess'.

a. If a day of the week is picked at random what is the probability that it is Monday?

Solution   *1 in 7 (or $\frac{1}{7}$).*

b. What is the probability of snow on Christmas Day? Give reasons.

Solution   *Any answer supported by a sensible explanation would be correct.*

List all of the possible ways in which something can happen, eg all the possible ways in which two dice can land.

Be able to list all of the different ways in which an event can happen. Eg if Adam, Barry and Carl are in a race the possible results are:

First	Second	Third
Adam	Barry	Carl
Adam	Carl	Barry
Barry	Adam	Carl
Barry	Carl	Adam
Carl	Adam	Barry
Carl	Barry	Adam

You need to develop a clear method when presenting results. Note how the results are given for the race. Can you see the logical approach?

Level 4 – Handling data

List the different ways of obtaining two heads and two tails when tossing four coins.

Solution
H	H	T	T
H	T	H	T
H	T	T	H
T	H	H	T
T	H	T	H
T	T	H	H

Note the logical way in which the solution is presented.

Have an understanding of median and mode.

a. Know that the median is the middle number when the numbers are in order.

b. Know that the mode is the most common number.

We frequently use the word average in everyday life, but average can mean many different things. Two mathematical names for types of average are median and mode.

Find the median and mode of these numbers,

1, 7, 5, 3, 2, 5, 3, 6, 5, 1, 2

Method   First place the numbers in order.

1, 1, 2, 2, 3, 3, 5, 5, 5, 6, 7

Solution   The median is the middle number, ie 3.

The mode is the most common number, ie 5.

Level 4 – Handling data

Attainment Test 4/6 *Answers on page 144*

1. These are the shoe sizes of twenty people.

    3   4   6   5   7   4   7   5   4   6
    4   3   6   7   5   5   4   5   6   6

    (a) Copy and record this data in the table below.

Shoe size	Tally	Frequency
3		
4		
5		
6		
7		

   (b) Find the arithmetic mean (average) of this data.
   (c) What is the range of this data?
   (d) Draw a frequency graph to show the information in the table above.

2. Draw a decision-tree diagram to sort out the pupils in a class into these four categories. Boys who take sandwiches, boys who do not take sandwiches, girls who take sandwiches and girls who do not take sandwiches for school lunch.

3. This graph shows the number of customers who visited a record shop in the first week in July.

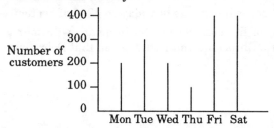

   (a) How many customers visited the shop on Monday?
   (b) What was the total number of customers for the week?
   (c) What was the mean (average) number of customers per day?
   (d) On which days did the shop have more than the average number of customers?
   (e) Why do you think the shop had more customers on Friday and Saturday?

Level 4 – Handling data

4.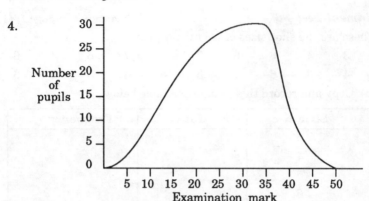

Use this graph to answer the following questions.
(a) How many pupils scored 15 marks?
(b) What was the highest mark scored?
(c) What was the most common mark?

5. Make a reasonable assessment of the following events happening. Choose your answer from
   No chance, Poor chance, Even chance, Good chance, Certain.
   (a) Sunday following Saturday.
   (b) Scoring an even number on a die.
   (c) Snow in September.
   (d) Mr Jones having his birthday on 32nd of September.

6. List all of the possible results in a race with four girls, Aisha, Barbara, Chrisoulla and Debra, given that Debra was last.

Level 5 – Handling data

> Design and use a simple recording sheet, eg use the sheet to record the colours of cars passing the school. Analyse the information collected.

Be able to conduct a simple survey without assistance.

> You will be given a task to complete and then expected to choose a suitable method for collecting and presenting the data collected.

A traffic survey is needed to find the number of road users.

i. Design a method for collecting data.
ii. Present the data in a suitable way and analyse your findings.

> Collect information and place the information into categories or groups.

Be able to gather, present and analyse data. (This is a frequency table for grouped data.)

Age of pupil (A)	Frequency (ie number of pupils)
$5 \leq A < 8$	87
$8 \leq A < 11$	89
$11 \leq A < 14$	97
$14 \leq A < 17$	83

> This will help to develop your analytical skills.
> *Note* The information in the frequency table in box 2 is in equal class intervals. If the ages were $5 \leq A < 8$, $8 \leq A < 9$, $9 \leq A < 17$ this would be unequal class intervals.

You would be required to present data which you had collected in a frequency table such as that shown above.

> Put information into a computer and make inferences based upon the information displayed by the computer.

a. Be able to extract information from a computer database, ie the contents or information held in the computer's memory bank.
b. Be able to analyse the information extracted from the computer.

Level 5 – Handling data

a. This helps to develop skills in using the computer.
b. This helps to develop your ability to 'think'.

Draw and make use of pie charts.

a. Be able to draw a pie chart.

Method   Draw a circle with a pair of compasses and use a protractor to measure the angles at the centre of the circle.

Be able to calculate the angle at the centre of the circle.
(***Note*** Angle at the centre of a circle equals 360°)

b. Understand and extract information from pie charts. Draw conclusions from the findings.

This is a pie chart to show categories of passengers travelling on a plane.

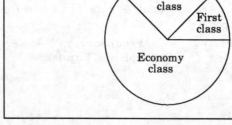

a. Draw a pie chart to show the way pupils travelled to school. 50 pupils arrived by bicycle, 70 arrived by bus, 60 arrived by car.

   Solution   *The angles at the centre of the circle should be Bicycle 100°, Bus 140°, Car 120°.*

b. Using the pie chart in the box above and given that there were 180 passengers on the plane, how many passengers travelled First Class?

   Solution   *24 First Class passengers.*

Draw and make use of conversion graphs.

Be able to draw a conversion graph.

Level 5 – Handling data

Example of a conversion graph.

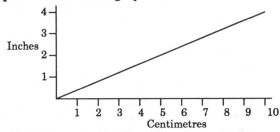

Conversion graphs are frequently used to convert from one scale (or unit) to another. The conversion graph in the box above converts centimetres to inches (and vice versa).

The rate of exchange is £1 = $1.60. Draw a conversion graph to show this and use your graph to find the £ value of $4.
Solution   £2.50.

Draw and make use of frequency diagrams for continuous variables.

Know the difference between continuous and discrete variables. (See the parrot box on page 93.) You should be able to use the inequality signs $>$ $<$ $\geq$ $\leq$ (ie greater than, less than, greater than or equal, less than or equal) for continuous variables.

This is an extension of the graphical work at **Level 4**. The difference is that you need to handle a continuous variable. The ability to read and understand graphs is very important.

Which of these graphs is for a discrete variable and which is for a continuous variable?

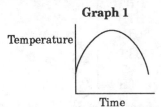
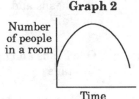

Solution   *Graph 1 in a continuous variable. (Temperature can have any value, eg 8.293°C.) Graph 2 is a discrete variable. (There can only be whole numbers of people.)*

Level 5 – Handling data

Know that probability experiments do not always produce the same results.

Understand that there are no certainties when dealing with probabilities.

The probability of a die landing on a '3' is $\frac{1}{6}$ (1 in 6). But if the die is shaken six times there is no guarantee that a '3' will occur exactly once. However if the die is shaken 6 000 times we would expect *about* 1 000 threes.

In five experiments cartons of eggs, each containing ten eggs, were dropped from the roof of a building. The numbers breaking were 8, 6, 7, 8, 7. If a sixth box were dropped how many eggs would you expect to break?

Solution   *If we take the mean average this would suggest 7.2 eggs should break, therefore 7 eggs would be a reasonable estimate. Also an answer of between 6 and 8 eggs makes sense, but there is no guarantee. Next time, no eggs may break, or all ten eggs may break.*

Know that probability estimates can be based upon repeated observation of an event or upon mathematically proven results based upon symmetry.

a. Know that probability can be based upon observation.

b. Know that probability can be based upon statistical fact associated with symmetry.

Example When three coins are tossed the possibilities are:

Three heads	HHH	$\frac{1}{8}$
Two heads and one tail	HHT HTH THH	$\frac{3}{8}$
One heads and two tails	HTT THT TTH	$\frac{3}{8}$
Three tails	TTT	$\frac{1}{8}$

Level 5 – Handling data

a. Probability can be based upon statistics. If an event is observed for a period of time, eg suppose there was an average of 11 days rain in September over the last 10 years, this would suggest that the probability of rain days this September is $\frac{11}{30}$.

b. Look at the coins table on page 102. The probabilities when tossing coins are based upon symmetry. Notice the chance of three heads is the same as the chance of three tails. The chance of two heads is the same as the chance of two tails.

Note the total probability is

$$\frac{1}{8} + \frac{3}{8} + \frac{3}{8} + \frac{1}{8} = 1$$

a. The probability of a man wearing glasses is $\frac{1}{10}$. How many men would you expect to wear glasses in a hall containing 250 men?

Solution   25.

b. Construct a table like the one on page 102 for four coins and then state the probabilities of each of the following.

   i. Four heads.
   ii. Three heads and one tail.
   iii. Two heads and two tails.
   iv. Three tails and one head.
   v. Four tails.

Solution   i. $\frac{1}{16}$ ii. $\frac{1}{4}$ iii. $\frac{3}{8}$ iv. $\frac{1}{4}$ v. $\frac{1}{16}$.

Know that if there are a number of events, say 5, and each event is equally likely, the probability of each event is

$$\frac{1}{\text{Number of events}}$$

in this case $\frac{1}{5}$.

Should have a good knowledge of basic probability.

103

Level 5 – Handling data

> If a certain event can occur in a number of ways – eg three names, Anita, Barbara and Carol, are placed in a hat – the probability of each name being selected is equal. The total number of possible selections is three. Therefore the probability of each name being selected is $\frac{1}{3}$.
> ***Note*** $\frac{1}{3} + \frac{1}{3} + \frac{1}{3} = 1$

What is the probability of a die landing on a '5'.

Solution $\frac{1}{6}$. *(There are six sides, only one of which is a '5'.)*

Level 5 – Handling data

Attainment Test 5/8 *Answers on page 148*

1. These are the heights of twenty pupils in metres.

   1.73 1.46 1.50 1.63 1.54 1.62 1.53 1.49 1.83 1.72
   1.64 1.90 1.73 1.52 1.60 1.73 1.52 1.49 1.57 1.62

   (a) Copy and complete this table.

Height in metres	Tally	Frequency
1.4 m and less than 1.5 m		
1.5 m and less than 1.6 m		
1.6 m and less than 1.7 m		
1.7 m and less than 1.8 m		
1.8 m and less than 1.9 m		
1.9 m and less than 2.0 m		

   (b) Draw a frequency graph to show this information.

2. This pie chart shows the lunchtime arrangements for 720 pupils in a school.

   (a) What is the value of x?
   (b) How many pupils take school lunch?
   (c) How many pupils take sandwiches?

3. 120 people were asked which television channel was their favourite.

   42 stated BBC1
   23 stated BBC2
   44 stated ITV

   The rest stated Channel 4.

   (a) Copy and complete this table in order to draw a pie chart.

Favourite channel	Number of people	Angle at the centre of the pie chart
BBC1		
BBC2		
ITV		
Channel 4		
Total		

   (b) Show this information on a pie chart.

4. The exchange rate is 3 German marks to the pound (£).

   (a) Draw a conversion graph to show this information.

Level 5 – Handling data

    (b) Use your graph to answer the following questions:
        (i) How many pounds (£'s) would I receive for 4.5 German marks?
        (ii) How many German marks would I receive for £2.50?

5. These are the weights, in kilograms, of the pupils on a bus.
   35.3 37.2 40.6 41.2 39.7 43.4 48.9 37.8 43.4 49.3
   37.6 42.3 37.3 40.2 36.0 38.2 48.0 39.6 38.3 42.0
   Choose three equal class intervals for the weights and present the data in a table like this.

Weight	Tally	Frequency

6. The probability of a train arriving late at the station is 1/8, the probability of a train arriving early is 1/10.
   (a) On Wednesday 240 trains arrived at the station. How many would you expect to be late?
   (b) On Thursday 35 trains arrived late. How many trains would you expect to arrive early?

7. (a) Complete this table to show all of the possible ways in which two dice can land.

       Total score
       2   1 + 1
       3   1 + 2, 2 + 1
       4   1 + 3, 2 + 2, 3 + 1
       etc.

   (b) What is the probability of the total of two dice being exactly 5?
   (c) What total is the most likely to occur?
   (d) What is the probability of shaking a double?

8. David is a member of a soccer team of eleven players. A boy is to be selected as the captain. The names of all the players are placed in a hat. What is the probability that David will be selected as captain?

Level 6 – Handling data

> Decide upon a situation for which information can be collected using a simple recording sheet. Analyse the information collected.

Be able to recognise a problem which would lend itself to mathematical analysis.

> You need to be aware of a variety of data collection methods and methods of presenting data such as tables, graphs, etc.

This could include a wide variety of tasks, eg how could you alter the parking bays in the school car park to allow more cars to park, or study the dinner queue to find a more efficient system?

> Conduct a survey to find out the opinions of people on some issue. Present the information and comment upon the results.

Be able to carry out a complete survey from start to finish and understand the effect of bias on the results. (Bias is caused by an unrepresentative sample of people being chosen for interview. Poor samples can cause misleading results, eg if everyone interviewed was under the age of 18 the results might not be representative of people over 18.)

> In these tasks you are expected to make comments about the data collection methods and display and comment upon the findings and results.

Carry out a survey to find opinions on the question of whether shops should be allowed to open on Sundays. *Note* account would need to be taken of the bias, eg shop workers would have different opinions to customers.

> Present information in scatter graphs.

Be able to draw a scatter graph and to interpret the results, ie understand the types of correlation.

Level 6 – Handling data

Positive Correlation

A pupil with a high mark in English is likely to gain a high mark in Maths.

Negative Correlation

The lower the number of cigarettes smoked the more likely a person will be to live longer.

No Correlation

The last digit of the house number is not correlated (has no connection) with the last digit of the telephone number.

Scatter graphs are useful for showing correlation (this means a connection) between data. For example you would expect that the more hours a person studies, the better grades he or she would obtain.

Draw a scatter graph to show the ages of pupils along one axis and the heights of the pupils along the other. Comment on your results.

Solution — *We would expect positive correlation, as pupils become older they become taller.*

Present information in two-way tables and understand the information presented in two-way tables.

Be able to make, complete and analyse the information presented in a two-way table.

Example

Features \ House	1	2	3	4	5	6
Garage	✓		✓	✓	✓	
Bedrooms	3	4	4	3	3	4
Garden	✓	✓		✓	✓	
Study				✓		✓

People are often required to examine information and then take action on their findings, eg selecting a suitable home – a list showing all of the desired features can then be completed. (See example above.)

Level 6 – Handling data

Ask your friends to state the number of hours they spent doing homework last night. Record the information, analyse it and comment upon possible bias.

Possible solution

Hours	0	1	2	3	4
Adam		✓			
Brian				✓	
Claire			✓		
Debbie					✓

The average homework time could be found, who worked longest, etc.

Bias Are your friends telling the truth? Is there proof that they spent the number of hours they claim?

Draw and understand network diagrams.

Be able to draw, understand and use a network diagram. This example of a network diagram shows the bus and train routes between five towns.

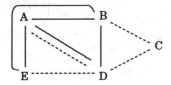
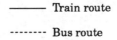
——— Train route
------- Bus route

Information can be presented in a wide variety of ways. This network diagram allows us to see at a glance the bus and train routes between towns. The same information could be presented in a table as shown below.

From \ To	Train routes				
	A	B	C	D	E
A		✓		✓	✓
B	✓			✓	
C					
D	✓	✓			
E	✓				

You are expected to be able to construct the networks from tables, and tables from networks.

Level 6 – Handling data

Use the network diagram in the box above to answer these questions.

a. Is it possible to get from B to E by train only?
b. Is it possible to get from B to E by bus only?
c. One of the towns does not have a railway station. Which town is this?

    Solutions   *a. Yes: B to A to E*
                         b. Yes: B to C to D to E
                         c. Town C.

> Be able to list all of the possible results for two independent, combined events. Be able to present the results in a diagram or table.

Know how to present information in diagrammatic or tabular form.

Diagrammatic form   Tree diagram to show the outcome of tossing three coins.

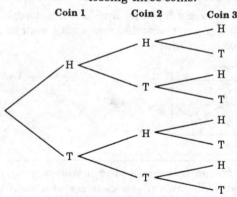

Tabular form   See *Solution* below.

> Independent events are events which are not affected by each other, in the example in the box above, the outcome of tossing the second coin is not affected by the outcome of tossing the first coin. In the example we combine the three outcomes to produce a combined event, ie the outcome of tossing three coins.

List all of the possible outcomes when tossing three coins.
    Solution   *Using the tree diagram above we have:*
                      HHH, HHT, HTH, HTT, THH, THT, TTH, TTT
(This is an example of presenting results in tabular form.)

Level 6 – Handling data

> Know what exhaustive and mutually exclusive events are.
> Know that the total probability for an event is 1. Know
> that the probability of an event occurring + the probability
> of an event not occurring equals 1.

a. Know that the total possible outcomes for any event is 1.
b. Know that the probability of an event happening plus the probability of it not happening equals 1.

Example

The probability of not throwing a 5 with a die is 1 minus probability of throwing a 5.

$$\text{ie } 1 - \frac{1}{6} = \frac{5}{6}$$

> a. An example of an exhaustive and mutually exclusive event would be the tossing of a coin. It can land on a head or a tail and nothing else. If it were possible for the coin to land on its edge, then the result head-tail would not be exhaustive. It is mutually exclusive because a coin which is only tossed once, if it lands on heads, cannot land on tails in that toss.
>
> b. This question is often asked: 'What is the probability of obtaining at least one tail when tossing four coins?'. This means exactly the same as 'What is the probability of not obtaining four heads when tossing four coins?'.
>
> *(Solution Probability of four heads is $\frac{1}{16}$. Probability of at least one tail is $1 - \frac{1}{16} = \frac{15}{16}$.)*

a. Make a list of all of the possible outcomes when tossing two coins. From this list state the probability of tossing two heads.

Solution   HH HT TH TT

The probability of each event is equal, ie $\frac{1}{4}$.

The probability of tossing two heads is $\frac{1}{4}$.

b. When tossing two coins what is the probability of not tossing any heads.

Solution $\frac{1}{4}$ (ie $1 - \frac{3}{4}$).

Level 6 – Handling data

Attainment Test 6/5
Answers on page 152

1. Describe the type of correlation shown by these scatter graphs.

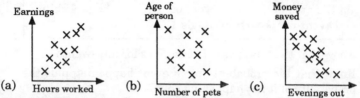

2. Draw a scatter diagram to show the information in this table.

Candidate	A	B	C	D	E	F	G	H	I	J
Maths mark	5	4	7	5	8	3	2	9	4	6
English mark	6	5	5	4	7	2	2	8	4	5

Describe the correlation shown by your scatter graph.

3. Show this information in a two-way table.

    Hours of television watched Adam 5 hours, Brendan 2 hours, Carolyn 4 hours, Deborah 3 hours, Elaine 1 hour, Francis 6 hours, Gordon 4 hours, Hannah 3 hours.

4. This table shows the hire purchase repayments for buying goods from a large store.

Money borrowed (£'s)	Weekly repayment over one year (£'s)	Weekly repayment over two years (£'s)
10.00	0.24	0.13
20.00	0.48	0.26
30.00	0.72	0.39
40.00	0.96	0.52
50.00	1.20	0.65
100.00	2.40	1.30
200.00	4.80	2.60
300.00	7.20	3.90

   (a) What is the weekly repayment over a period of two years if you buy a coat for £30?

   (b) Mr White bought a suit. His weekly repayment was £1.68 for a period of one year. What was the cost of the suit?

   (c) Mrs Jones bought a washing machine costing £340. She pays the money back over 2 years.

   (i) What is her weekly repayment?

   (ii) How much money would she save if she paid cash for the washing machine?

Level 6 – Handling data

5. Four people A, B, C and D were asked to write down the names of the people in their group who they thought were sensible. The results are shown in this network. An arrow from A to B means that A thought B was a sensible person.

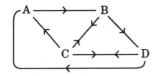

Copy and complete this table. (The first row has been completed for you.)

	A	B	C	D
A	0	1	0	0
B				
C				
D				

6. This table shows the direct bus routes in town.

    (a) Draw a network diagram to show the information in the table.

    (b) From the network diagram explain how you would get from A to D.

		\multicolumn{4}{c}{To}
		A	B	C	D
From	A		✓	✓	
	B	✓			✓
	C		✓		
	D	✓		✓	

7. Draw a tree diagram to show the outcome of tossing four coins. From the diagram determine:

    (a) The probability of tossing exactly two heads.

    (b) The probability of tossing at least two heads.

8. A bag contains 4 red sweets, 5 blue sweets and 1 yellow sweet.

    (a) What is the probability of choosing

    (i) a red sweet;

    (ii) a blue sweet;

    (iii) a yellow sweet?

    (b) What is the probability of not choosing a red sweet?

Level 7 – Handling data

Choose a situation which is suitable for investigation. Make a guess at what the result of the investigation will be. Produce a yes/no questionnaire to test the opinions of people. Examine the results and state whether the original guess was right or wrong.

Be able to think of an idea which can be tested using a questionnaire, eg test the statement (or hypothesis) 'Pupils at school would like to have a sweet shop'.

A hypothesis is a statement which has not been proved to be correct or false. The hypothesis in the box above 'Pupils at school would like to have a sweet shop' could be tested by asking all of the pupils in you school for their answer. If the majority agreed then the hypothesis would be valid.

Think of a statement which can be proved or disproved (ie a hypothesis). Design an appropriate questionnaire to test whether your hypothesis is valid.

Collect statistics, place these statistics into groups, use a variety of techniques to analyse the data.

a. Know how to organise and present data.
b. Know how to group data into class intervals, eg data, such as the distance pupils travel to school, can be collected and then presented in a table as shown. D is the distance in kilometres.

Class interval	Mid interval point	Frequency
$0 < D \leq 2$	1	12
$2 < D \leq 4$	3	8
$4 < D \leq 6$	5	6
$6 < D \leq 8$	7	3
$8 < D \leq 10$	9	1

c. Understand the meaning of frequency.
d. Know how to calculate the range of a set of data.
e. Know how to calculate the mean of a set of data.
f. Know how to calculate the mode of a set of data.
g. Know how to calculate the median of a set of data.
h. Be able to use the mean, median, mode and range of sets of data to make comparisons.

Level 7 – Handling data

a. You should be able to use all of the data presentation methods already described.

b. Note that the class intervals in the box above are all equal, ie 2 km. It is always advisable to keep the class intervals equal. The mid-point is the mid point of the class interval, eg mid-point of $2 < D \leq 4$ is 3. The mid-point is used when calculating the mean.

c. The table in the box above shows that 8 pupils travelled between 2 and 4 kilometres. 8 is the frequency.

d. The range is the difference between the smallest and largest, eg from 0 to 10. A range of 10.

e. The mean for a set of data can be calculated as follows (from the table in the box above)

$$\frac{1 \times 12 + 3 \times 8 + 5 \times 6 + 7 \times 3 + 9 \times 1}{30}$$

The mean distance travelled is 3.2 km.

f. The mode (or modal group) is the most common group. In the example most pupils travel between 0 and 2 kilometres (ie the mode).

g. The median is the middle value when the data is in order (ie smallest to largest). The median for the distance to school is the distance travelled by the $\frac{15}{16}$ pupil, ie between 2 and 4 km.

h. It is easier to compare data when we have the mean, median, mode and range of the data.

i. You would be required to present data and analyse the results using all of the techniques shown in the box above.

ii. Use the following sets of data to calculate:

   1. the range; 2. the mean; 3. the median; 4. the mode.

   W = the weight of men in kilograms.

Class interval	Mid-point	Frequency
$40 < W \leq 50$	45	3
$50 < W \leq 60$	55	7
$60 < W \leq 70$	65	22
$70 < W \leq 80$	75	38
$80 < W \leq 90$	85	22
$90 < W \leq 100$	95	8

Solutions

1. The range of the data is 40 to 100, ie a range of 60.

Level 7 – Handling data

2. The mean is 74.3 kg.
3. The median is the weight of the middle person, ie man 50/51. This is the class interval $70 < W \leq 80$.
4. The mode or modal group is $70 < W \leq 80$ because there are more men of this weight than any other group.

Note There are a variety of more complex methods and graphical methods which can be used. However, these methods are not within the scope of this book.

Draw a frequency polygon and line graph from information presented in tables. Know how to compare two graphs.

a. Be able to draw a line graph and a frequency polygon for a frequency distribution.

Example This table shows the ages of people in a village.

Ages	Mid-point	Frequency
$0 < A \leq 10$	5	6
$10 < A \leq 20$	15	5
$20 < A \leq 30$	25	7
$30 < A \leq 40$	35	6
$40 < A \leq 50$	45	3
$50 < A \leq 60$	55	4
$60 < A \leq 70$	65	2

The information in the table shown in a

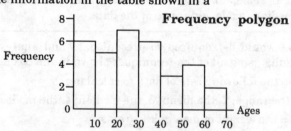

The information in the table shown in a

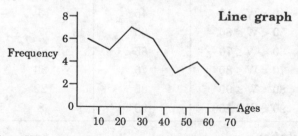

Note When drawing a line graph it is the mid-points which are joined.

Level 7 – Handling data

b. Be able to compare two frequency distributions.

   Example This table shows the English and Maths marks for 106 pupils.

Class interval (Marks)	Mid-point	Frequency English	Frequency Maths
1 – 20	10.5	5	15
21 – 40	30.5	13	35
41 – 60	50.5	34	38
61 – 80	70.5	42	11
81 – 100	90.5	12	7

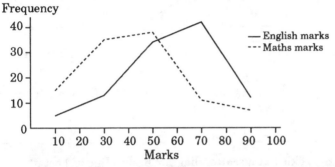

- a. Graphs can be used to provide a clear visual display of mathematical information.
- b. The graph shows that the English marks are generally higher than the maths marks.

a. You will be required to draw graphs from information presented in tables.

b. You will be required to interpret the meanings of the information presented in graphs. You may be asked to draw inferences from the information provided in graphical form.

Draw and understand flow diagrams (often used in computing).

Be able to construct a flow chart.

A loop in a flow diagram has the effect of repeating a sequence of instructions until a given condition is met. See the solution in the box below for an example.

Level 7 – Handling data

Construct a flow diagram to find the lowest multiple of 6 which is greater than 32.

Solution

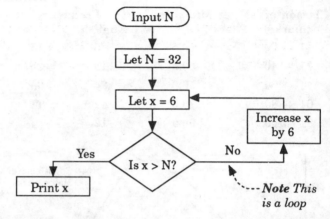

The number which will be printed is 36. (There are a variety of other correct programs.)

By looking at a scatter diagram draw a line of 'best fit'.

Be able to draw a scatter diagram (see page 108) and estimate the location of a line of 'best fit'.

A line of 'best fit' is drawn by looking at the crosses on a scatter diagram and assessing the best place to position the line. Normally there would be a similar number of crosses above the line as below. A line which slopes upwards shows positive correlation between the two variables, a line which slopes downwards shows negative correlation between the two variables. (See page 108.)

Place a line of best fit on this scatter graph.

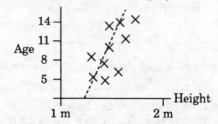

Solution *The line of 'best fit' is shown by the dotted line above.*

Level 7 – Handling data

> Be able to estimate probability.

a. Be able to observe an event for which the probability is unknown, establish the frequency of the event, and produce an estimate of the probability of the event.

b. Be able to observe an event for which the probability is known and confirm that the known probability is correct. Eg we know that a coin will land on heads when tossed about half the time. If we test this assumption by experiment this should be proved. (A high number of tosses would be needed.)

> a. By observing an event for a sufficiently long period of time it is possible to estimate the frequency of that event.
>
> b. Know that the more times an experiment is repeated, the nearer we would expect the actual number of events to equal the estimated number of events, eg if we throw a die thousands of times we would expect each of the numbers to appear about one-sixth of the time.

a. It has snowed on 20 Christmas days during the last 200 years. Based upon this information what do you think the probability of snow on Christmas day next year will be?

Solution $\frac{20}{200} = \frac{1}{10}$ (or 0.1).

b. A die was rolled 300 times. The results were as follows:

1 – 18 times; 2 – 52 times; 3 – 48 times;
4 – 57 times; 5 – 44 times; 6 – 81 times.

Comment upon these results.

Solution *Something appears to be wrong. We would expect each number to occur about one-sixth of the time, ie 50 times. This has not happened. Because the 6 is opposite the 1 on a die, the results suggest that the die may not be fair, ie it is weighted. More rolls of the die are needed. If after 3 000 rolls the same pattern continued, we could state that the die was weighted.*

Level 7 – Handling data

Know that some probabilities cannot be estimated because there is insufficient evidence available. Therefore an 'educated guess' has to be made. Eg 'What is the probability of a man landing on Saturn?' We can only guess. But we know that the chance of a die landing on '3' is about one in six.

Know that when assigning a probability to an event, guesses sometimes have to be made. Past evidence may not be appropriate, eg in 1965 one train in every three was a steam train. This probability would no longer be the case in 1992.

Subjective estimates are really no more than intelligent guesses. They are based upon a person's assessment of the situation and not upon 'hard facts'.

You will be required to decide if probabilities are based on evidence or are subjective estimates.

Know how to calculate the probabilities for mutually exclusive events.

Know that the probabilities for mutually exclusive events can be calculated by adding the individual probabilities.

'Mutually exclusive events' are events which cannot both happen, eg if we throw a die we can throw a 3 or a 4, but we cannot throw a 3 and a 4 at the same time.

What is the probability of shaking a 6 or an odd number on a die?

Solution Probability of an odd number = $\frac{1}{2}$

Probability of a 6 = $\frac{1}{6}$

Total probability = $\frac{1}{2} + \frac{1}{6} = \frac{2}{3}$.

Level 7 – Handling data

Attainment Test 7/6   *Answers on page 155*

1. These are the heights in metres of boys in the third year.

    1.403 1.690 1.520 1.634 1.522 1.531 1.602 1.482 1.673 1.584
    1.682 1.528 1.470 1.524 1.670 1.600 1.550 1.620 1.430 1.570

    (a) Choose six equal class intervals, copy and complete this table. H is the height in metres.

Height	Tally	Frequency	Mid-point
$1.40 \leq H < 1.45$			
$1.45 \leq H < 1.50$			

    (b) Calculate the mean height of the pupils using the 'mid-point' method.

2. These are the test marks of eleven pupils.

    2, 7, 3, 9, 4, 8, 7, 7, 4, 6, 5

    (a) What is the range of the marks?
    (b) What is the mode?
    (c) What is the median?
    (d) What is the mean?

3. This table shows the ages of men and women at a disco.

Age	17	18	19	20	21	22	23	24	25
Men	5	8	18	25	35	40	35	17	8
Women	15	20	31	37	30	26	18	5	1

    (a) Draw a line graph to show the frequency distribution for the men.
    (b) On the same graph, but in a different colour, show the frequency distribution for the women.
    (c) Compare the two frequency distributions and make comments on your observations.

4. Construct a flow diagram to find the lowest square number over 300.

5. This scatter diagram shows the number of days pupils were absent from school and their ages.

    Draw a line of 'best fit' by inspection on the scatter diagram and state the type of correlation shown.

Level 7 – Handling data

6. A bag contains 3 red sweets, 4 blue sweets, 5 green sweets and 6 white sweets. What is the probability of choosing:
   (a) a red sweet;
   (b) a red sweet or a blue sweet;
   (c) a red or a blue or a green sweet;
   (d) a red or a blue or a green or a white sweet;
   (e) a yellow sweet?

Using and Applying Mathematics

This section of the National Curriculum consists of using the mathematics which has been acquired in the other parts of the National Curriculum, ie material in the topics of this book.

This section will be tested during normal lessons as part of the teacher assessment. It will *not* be tested in the official DES tests at the end of Key Stage 3.

Using and applying mathematics

Section A – Using mathematics

Level 4 | Choose the mathematics needed for a project. Plan the work.

You will be given a task, you will then be expected to decide upon a suitable approach, plan the work and decide what equipment and mathematical knowledge you will need.

Level 5 | Choose the mathematics needed for a project, use an orderly procedure and check the development of the work at regular intervals.

Level 5 builds on Level 4. You will be expected to decide what information is needed to tackle the project. You will also be expected to consider the progress made at several stages of the project and the implications this might have for future approach to the project. You should be able to determine whether there is enough information to complete the project.

Level 6 | Decide upon a mathematical project or investigation. Find ways of obtaining the required data. Use 'trial and improvement' methods if applicable.

Level 6 builds on the previous levels. You will be considered capable of choosing a mathematical project of your own. You would be expected to use more advanced techniques and approaches to the project. You should be able to identify any information which is required, but is not available, and then choose a method to obtain the information, eg in studying traffic patterns a survey may be required to obtain the information.

You will be expected to understand how to use 'trial and improvement' methods. This means find a solution, which may not be the best solution. This solution is then tested: from this it may be possible to improve the solution.

Level 7 | Decide upon a mathematical project or investigation and use an orderly procedure to conduct the investigation. Be selective in the use of information. Check the development of the work at regular intervals, using 'trial and improvement' methods if applicable.

Level 7 builds on the previous levels. At this level you should be able to structure your own work. Decide which direction the investigation

Using and applying mathematics

should take. Decide which information is useful and which is not needed and should be discarded.

Section B – Mathematics communication

Level 4 | **Display information in a variety of forms.**

You should be able to communicate your ideas and findings in spoken language and in writing. You should also be able to draw simple line graphs and bar charts to show your findings.

Level 5 | **Make simple inferences about information presented in a variety of forms.**

Level 5 builds on Level 4. You should be able to read, understand and make comments about information provided in a variety of ways – eg graphs, tables, pie charts and pictures.

Level 6 | **Make use of and be able to display mathematical information in a variety of forms.**

Level 6 is very similar to Level 4. However more sophisticated techniques should be expected. The written explanations should be clearer. The visual form should include more complex graphical displays such as scatter diagrams and pie charts.

Level 7 | **No specific description of this level is given.**

This level is not described in the National Curriculum. Therefore the requirements are the same as for Level 6.

Section C – Mathematical reasoning

Level 4 | **Use numbers to test simple mathematical statements.**

You should be able to check that your mathematical statements are correct by using actual numbers and determining whether the results are sensible and accurate.

Using and applying mathematics

Level 5	**Make simple mathematical statements and use numbers to test them.**

Level 5 is an extension of Level 4. In addition to testing statements you would also be expected to make statements which can be mathematically tested, eg you might make the statement 'more women than men go to supermarkets'. The statement could then be tested by a survey to count the number of men and women.

Level 6	**Make and test simple hypotheses and generalisations.**

You should be able to produce and check generalisations and simple hypotheses.

A generalisation means a rule. A hypothesis means an idea.

Example The hypothesis 'If you shake a fair six-sided die, it will land on three about one-sixth of the times' might be made. The hypothesis could be tested by experiment. Shake the die 600 times, record the results. We would expect the number of threes to be about 100. From this a generalisation could be produced.

Probability of an event

$$= \frac{\text{Number of ways in which the desired result can occur}}{\text{Total number of ways in which the event can occur}}$$

$$= \frac{\text{One way of shaking a three}}{\text{Six possible ways of a die landing}}$$

$$= \frac{1}{6}$$

Level 7	**Understand mathematical explanations and identify errors in mathematical information.**

Level 7 is an extension of the previous levels. You will be expected to follow a textbook explanation such as the type which occur at the start of chapters.

As you develop in mathematical ability you will be expected to produce your own explanations for observed mathematical results and sequences. You should be able to test your theories and identify steps in the reasoning which do not make sense, or do not apply in all cases. You should then be able to tackle the problem, project or investigation in a new way.

Using and applying mathematics

Section D – An example of problem-solving in a real everyday situation

The old system in banks and post offices was that each counter had its own queue.

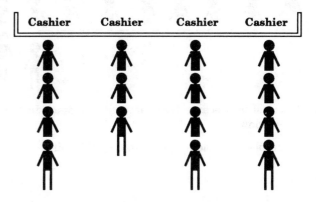

Each person would choose the queue which he or she thought was the shortest. The problem with this system was that some customers took much longer than others. Sometimes one person could take as long as five people in an adjoining queue.

Banks were faced with the problem of making the queuing process fairer.

This was a mathematical problem solving exercise. They studied the problem and produced a number of solutions to reduce queues.

1. They introduced 'cheque post'. Some customers were able to place their requests in an envelope and put them into the 'cheque post' box. This saved these people from queuing.
2. ATM cash card machines were introduced to allow people to obtain small amounts of cash without seeing a cashier.
3. A new type of queuing system was introduced: just one queue.

Using and applying mathematics

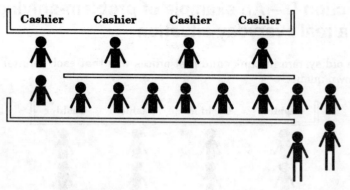

As soon as a cashier became free the next customer in the queue went to that window.

This is an example of problem solving in a real life situation.

> A variety of tasks for 'Using and Applying Mathematics' is shown in Sections E and F. The tasks are applicable to all levels of Key Stage 3 and it is the sophistication of the solutions and the methods used which determine whether the work is suitable for Level 4, 5, 6 or 7.

Section E – Practical tasks

Project 1
Test this hypothesis: 'The majority of car drivers are men'.

Project 2
Show, on an accurate scale diagram, the position of the furniture in your lounge. Can you find a better way of arranging the furniture?

Using and applying mathematics

Project 3

Study the school dinner queue, or any other queue. Look at the system. Can you find any way to improve the system?

Project 4

Study the way in which pupils move around the school at breaks, end of lessons, etc. Can you suggest improvements to the system to relieve congestion?

Project 5

Look at the design of the school car park or any other car park. Can you suggest methods of (a) improving access to parking bays (b) increasing the number of cars which the car park will hold? Would these changes have any disadvantages?

Project 6

Observe the pedestrian crossings in the High Street. Are they in the correct places? What effect would changes in their location have on (i) pedestrians (ii) traffic? (The project could be extended to consider the effect of the location of pedestrian crossings on trade in various shops.)

Project 7

Take ten drawing pins. Carry out a number of experiments to determine whether when dropped they are more likely to land point up or point down. Show your results and comment upon them.

Project 8

Take a paper tissue. Place a card 50 cm by 50 cm on the floor. Drop the tissue from various heights, say 20 times from a height of 50 cm, 20 times from 75 cm, 20 times from 100 cm etc. Record the number of times the tissue lands

(a) completely in the square;
(b) partly in the square;
(c) completely out of the square.

Present your results in a variety of ways. Construct a probability table for each height.

Using and applying mathematics

Section E: suggested solutions

Project 1

1. Carry out a survey. There are a variety of possible methods, e.g. stand by a road and count the number of male and female drivers. The results could then be shown as a percentage, in a pie chart, on a bar graph, etc.
2. Repeat the survey at a different time and in a different place, compare your results. Make comments upon your results. The more cars you can observe the more accurate your survey should be.
3. Consider the variables, eg does the time of day and the place where the survey is carried out affect the results. Consider the effect of carrying out the survey close to a school at the end of the school day or in a supermarket car park on a weekday morning. At such times the proportion of female drivers may well be greater.
4. Consider other problems with your data collection methods. Is counting the number of car drivers a true indication of the proportion of male and female drivers? If there is a woman in the passenger seat it is quite possible that she can also drive. Possibly a survey of one hundred men and one hundred women would give a better estimate of the proportion of male and female drivers.

Project 2

There are many ways of approaching this project. You should consider including the following in your project.

1. Scale drawing of your lounge.
2. Calculate the floor area.
3. Show the position of the doors and windows: which way do the doors open?
4. Television points may dictate the position of the television. A radius and arc could be shown on the plan for the best viewing distance.
5. Routes through the room will have to be indicated, and kept clear of furniture.
6. Different possible positions of furniture can be considered and shown, advantages and disadvantages can be discussed.

Project 3

The following could be included in your project.

Using and applying mathematics

1. A diagram and explanation of the existing system. A scale diagram showing the serving area, seating area and routes would be helpful. The entrance and exit should be shown.
2. Observe the queue. What causes hold-ups? Possibly design a questionnaire to ask pupils opinions of the system.
3. Time various activities, eg how long does it take to serve a drink, where is the hold-up, would a re-arrangement of serving staff help?
4. A graph or pie chart to show the meals selected. This may indicate a large number of people have cold meals – would two queues help, one for cold meals, one for hot meals?
5. Suggest changes, explain how they will help. Consider disadvantages of the changes as well as advantages.
6. Consider each year group having lunch at a different time.

Project 4

The following could be included in your project.

1. A map of the building.
2. Congestion points could be shown.
3. A survey could be carried out to show the number of pupils using each congested area, these are often entrances and exits.
4. This information could be shown in graphs, pie charts, tables, etc. It could also be shown as percentages.
5. Having identified the problem areas solutions should be suggested.
6. Consider the advantages and disadvantages of these changes.
7. Would it help to have all pupils keeping to the left when moving around the building? What about a one-way system, would that help?

Project 5

There are a variety of approaches to this project,. Much depends upon what problems the existing layout causes. What are your intentions: to provide as many parking spaces as possible, to make parking cars easy, or something else. The following could be considered.

1. Diagram to show the existing car park layout. Look at the advantages and disadvantages.
2. Calculate the size of the car parking bays, length, width and area.
3. Consider the turning circles of cars and the effect of having parking bays at an angle, rather than at right angles to the line of traffic. With a one-way system around the car park this might help.

Using and applying mathematics

4. Carry out surveys to identify who uses the car park.
5. Show the different sizes of vehicles which use the car park in a graph.
6. Consider possible changes, disadvantages as well as advantages of such changes.

Project 6

The following could be included in your project.

1. Diagram of the existing crossings and the High Street.
2. Survey of people to find their opinions.
3. Graphical presentation of the survey statistics – pie charts, graphs, etc.
4. Density of traffic, this could be shown in a table.
5. Do people use the crossings? If not, why not?
6. Does the positioning of the crossing have any effect on trade in the shops?
7. Are there any better positions for the crossings?
8. What effects do the crossings have on traffic patterns in the town? Do they cause congestion?

Project 7

1. The experiment should be carried out a number of times.
2. The results should be shown in graphs, charts, etc.
3. What is the effect of dropping the drawing pins from different heights?
4. Can you determine the probability of a drawing pin landing point up from different heights? Test your ideas.

Project 8

1. Produce probability tables for various heights.
2. Use pie charts and graphs to show your results.
3. How does the probability vary according to different heights?
4. Does the size of the tissue have any affect?
5. Is there any skill attached to dropping the tissue? Do the results improve with practice, or when different people drop the tissue?

Using and applying mathematics

Section F – Mathematical reasoning: tasks

Project 9

You start with eleven chairs, five boys and five girls. They sit as shown.

| B | B | B | B | B | | G | G | G | G | G |

The object is to move all of the boys to the right, the girls to the left. The final position is

| G | G | G | G | G | | B | B | B | B | B |

The rules

1. The boys are only allowed to move to the right, the girls to the left.
2. The children may either move into an empty seat beside them, or jump over one other person (as in draughts).

   Example

   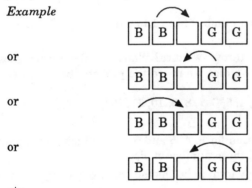

   etc.

3. If no-one can move according to the rules, the attempt has failed and everyone must return to their starting positions and try again.

(Counters may be used instead of people.)

Project 10

This is a well-known mathematical game for two players. Place the matchsticks as shown in the diagram.

The rules

First player removes as many matchsticks as he wishes, but only from one row, eg if he chose row 3 he could remove 1, 2 or 3 matchsticks. Players take it in turns to remove matchsticks. The player who removes the last matchstick loses.

Using and applying mathematics

Investigate this game. Try to find a winning method. Can you find a strategy to win? What are the winning positions with four matchsticks left, what are the losing positions with four matchsticks left?

Project 11

Place up to 40 cards or matchsticks on a table. (The smaller the number the easier it will be to find a winning strategy.)

This is a game for two players. Each player takes his turn to remove 1, 2 or 3 cards. The player who removes the last card wins.

a. Investigate the game. Decide upon a winning strategy. Sometimes it is better to go first, sometimes second. Explain why. Develop a strategy always to win if you can choose whether to go first or second.

b. What difference does it make if the rules are changes and players can remove 1, 2, 3 or 4 cards each turn?

Try to generalise, ie find a rule.

Project 12

If you throw one die each number from 1 to 6 has an equal chance of landing.

a. What happens if you throw two dice? Which number occurs most often? If you throw two dice what are the probabilities of each of these totals

   2, 3, 4, 5, 6, 7, 8, 9, 10, 11, 12?

   Is there a pattern?

b. Extend the investigation for 3, 4, 5 and 6 dice. Which total occurs most often for each of these number of dice?

c. What is the most likely total to be shaken when throwing:

   (i)   8 dice;

   (ii) 15 dice;

   (iii) 100 dice?

d. Generalise and determine which is the most likely total when n dice are shaken.

Using and applying mathematics

Section F: suggested solutions

Project 9

If you had problems with this project you should have used a well-tried technique. Try simplifying the problem. Find a solution for two boys and two girls. Then try three boys and three girls, etc. A solution for two boys and two girls is shown.

B	B		G	G
B		B	G	G
B	G	B		G
B	G	B	G	
B	G		G	B
	G	B	G	B
G		B	G	B
G	G	B		B
G	G		B	B

The solution for five boys and five girls is very similar.

Project 10

The sophistication of your solution would determine your attainment level.

At Level 4, you would be able to find a strategy to know whether you should go first or second to win when there are four matchsticks.

At Level 7, you could find a strategy to win with from the starting position, providing you could decide whether to go first or second.

Project 11

Because players can take a maximum of three cards the solution is based upon 4, ie one more than the maximum number of cards which can be taken.

Number of cards left	Next player	Number of cards left	Next player
1	Wins	6	Wins
2	Wins	7	Wins
3	Wins	8	Loses
4	Loses	9	Wins
5	Wins		

Using and applying mathematics

The table shows who wins providing both players make their best possible moves. The secret in this example is always to leave your opponent with a multiple of four cards, ie always try to leave 4, 8, 12, 16, 20, 24 etc, cards.

If 1, 2, 3 or 4 cards could be taken, then the solution is based upon 5. Always leave your opponent with a multiple of five cards to win.

Project 12

a. When throwing 2 dice there are 36 possible ways the dice can land.

Total	Ways of obtaining this total	Probability
2	1 + 1	$\frac{1}{36}$
3	1 + 2, 2 + 1	$\frac{2}{36}$
4	1 + 3, 2 + 2, 3 + 1	$\frac{3}{36}$
5	1 + 4, 2 + 3, 3 + 2, 4 + 1	$\frac{4}{36}$
6	1 + 5, 2 + 4, 3 + 3, 4 + 2, 5 + 1	$\frac{5}{36}$
7	1 + 6, 2 + 5, 3 + 4, 4 + 3, 5 + 2, 6 + 1	$\frac{6}{36}$
8	2 + 6, 3 + 5, 4 + 4, 5 + 3, 6 + 2	$\frac{5}{36}$
9	3 + 6, 4 + 5, 5 + 4, 6 + 3	$\frac{4}{36}$
10	4 + 6, 5 + 5, 6 + 4	$\frac{3}{36}$
11	5 + 6, 6 + 5	$\frac{2}{36}$
12	6 + 6	$\frac{1}{36}$

The pattern is shown above. The number which occurs most often is 7.

b. The numbers which occur most often when throwing 3 dice are 10 and 11.

The number which occurs most often when throwing 4 dice is 14.

The numbers which occur most often when throwing 5 dice are 17 and 18.

The number which occurs most often when throwing 6 dice is 21.

c. (i) 28 (ii) 52 and 53 (iii) 350.

d. To find the most common total for n dice multiply n by 3.5. If n is an even number this will give the most common total. If n is an odd number the result will end in 0.5: in this case take the number above and below, eg for 7 dice, $7 \times 3.5 = 24.5$, therefore the most common totals are 24 and 25.

Appendices

Appendix 1 – Mental arithmetic tests

Ask a friend to read these questions to you, write your answers on paper. You may not use a calculator. (Solutions on page 140.)

Time: Ten minutes for each test

Level 4   Test 4/7

1. It is exactly six weeks to Jayne's birthday. How many days is this?
2. Michelle travelled to work each day by bus. The return fare was 86 pence per day. What are her travel costs for five days?
3. What is the sum of the odd numbers below 10?
4. John had 93 sweets. He ate 27. How many sweets does he have left?
5. Write 8.20 pm using the twenty-four hour clock.
6. Sarah and Jayne earned £58 between them. They shared the money equally. How much did each receive?
7. Subtract 281 from 879.
8. Divide £56 equally between eight people. How much does each receive?
9. John weighs 33 kilograms. David is 9 kilograms heavier. How heavy is David?
10. Sarah is 5 cm shorter than her sister Anna. Anna is 138 cm tall. How tall is Sarah?

Level 5   Test 5/9

1. How many 14p stamps can be bought for £2.24?
2. A box contains 25 packets of chalk. Each packet contains 12 sticks of chalk. How many sticks of chalk are contained in one box?
3. A school had forty classrooms with thirty chairs in each classroom. How many chairs were there altogether?
4. John worked for thirty months and earned £800 per month. How much did he earn?
5. Sugar costs £0.34 per kilogram. What is the cost of 2.5 kilograms?
6. What is the product of 3 and 5?
7. Jayne wishes to buy a television costing £600. She saves £20 each week. How many weeks will she take to save enough money to buy the television?
8. How many times does 80 go into 4 000?
9. Mr Webb has 50 boxes of eggs. Each box contains 200 eggs. How many eggs does he have?
10. Janet buys 800 seeds. She decides to plant them in 40 rows. How many seeds should she plant in each row?

Appendix 1 – *Mental arithmetic tests*

Level 6   Test 6/6

1. How many minutes are in one and a quarter hours? 75
2. A train arrived at its destination at 14.25. The journey took fifty minutes. At what time did the journey start? 13.35
3. Write down the number three quarters of a million in figures. 750 000
4. Write down ten thousand and twenty in figures. 10 020
5. £50 was divided between David and Sarah in the ratio 3 : 2. How much money did each receive? 30 : 20
6. Mr Sawkins spent £3.72 at the supermarket. He paid with a £10 note. How much change should he receive? £6.28
7. What is 20% of £30? £6
8. Write 0.3 as a fraction. 3/10
9. A train travels a distance of 150 kilometres in two and a half hours. What is its speed in kilometres per hour? 60 km/h
10. I use three quarters of a packet of tea each month. How long will three packets of tea last me? 4 months

---

Level 7   Test 7/7

1. A room is 4 m long, 3 m wide and 2 m high. What is the volume of the room? 24 m³
2. Mrs Davis wishes to buy a car costing £8 000. She has twenty months to save the money. How much should she save each month? £400
3. How many bags of apples, each containing 0.6 kilograms, can be filled from 30 kilograms? 50
4. How many 2p coins are required to make £50? 2500
5. Divide 7 by 0.2.
6. A skyscraper has sixty floors. Each floor is 3 metres high. What is the height of the skyscraper? 180 m
7. Kevin lives 3.8 km from his school. The length of his pace is one metre. How many paces does he take to reach the school? 3800
8. How many litres of lemonade are contained in fifty 20 cl bottles?
9. 800 sheep are placed in pens. Each pen contains 25 sheep. How many pens are needed? 32
10. John has enough money to send thirty letters by second class post at 17p per stamp. Unfortunately he can only buy first class stamps at 22p per stamp. How many letters can he send? 23

Appendix 1 – Mental arithmetic tests

Solutions

One mark per question

Question	Level 4	Level 5	Level 6	Level 7
1	42	16	75	24 m^3
2	£4.30	300	13.35	£400
3	25	1 200	750 000	50
4	66	£24 000	10 020	2 500
5	20.20	£0.85	David £30 Sarah £20	35
6	£29	15	£6.28	180 m
7	598	30	£6	3 800
8	£7	50	$\frac{3}{10}$	10
9	42 kg	10 000	60 km/h	32
10	133 cm	20	4 months	23

Pass mark 8 out of 10 on each test.

Appendix 2 – Answers to attainment tests

[Marks allocated to each question are given in brackets]

Test 4/1

1. 1 001, 804, 739, 438, 192 *[1]*
2. (a) 304 *[1]*
   (b) 500 000 *[1]*
   (c) 5 027 *[1]*
   (d) 42 050 *[1]*
   (e) 107 000 *[1]*
3. (a) One hundred and eleven *[1]*
   (b) Three thousand, two hundred and eighty four *[1]*
   (c) Sixty thousand and twenty *[1]*
4. 225 000 *[1]*
5. 638, 893, 1 002, 3 204 *[1]*
6. (a) 320 *[1]*
   (b) 27 600 *[1]*
   (c) 4 520 *[1]*
   (d) 3 800 *[1]*
   (e) 7 000 *[1]*
7. £17 *[1]*
8. 68 kg *[1]*
9. (a) $\frac{1}{4}$ full *[1]*
   (b) $\frac{2}{3}$ full *[1]*
   (c) $\frac{1}{2}$ full *[1]*
   (d) $\frac{3}{4}$ full *[1]*
   (e) $\frac{1}{3}$ full *[1]*
10. Any ten sections shaded *[1]*
11. (a) 3% *[1]*
    (b) 40% *[1]*
    (c) 17% *[1]*
12. 16 *[1]*
13. 16% *[1]*
14. £1.68 *[1]*
15. £5.25 *[1]*

[Total = 31; Pass Mark = 21]

Test 4/2

1. 80 *[1]*

Appendix 2 – Answers to attainment tests

2. 3 500	*[1]*
3. £3.10	*[1]*
4. £2.31	*[1]*
5. 500 (*or* five hundred)	*[1]*
6. 68p	*[1]*
7. 62 boxes and 4p change	*[1]*
8. (c) 6 800	*[1]*
9. (c) 16 000	*[1]*
10. (d) 900	*[1]*
11. (a) 3	*[1]*
(b) 5	*[1]*
(c) 6	*[1]*
(d) 1	*[1]*
12. 20	*[1]*

[Total = 15; Pass Mark = 11]

Test 4/3

1. 7.8 cm	*[1]*
2. 8 700 g	*[1]*
3. 275.87 m	*[1]*
4. 28 800 seconds	*[2]*
5. Forty seconds after 10.46 pm on the 12th January	*[3]*
6. 5.21 pm (*or* 17.21)	*[1]*
7. 6 hours 15 minutes	*[1]*
8. (a) 10 mm	*[1]*
(b) 27 mm	*[1]*
9. (a) 1 cm	*[1]*
(b) 2.7 cm	*[1]*

[Total = 14; Pass Mark = 10]

Test 4/4

1. $3 \times 5 = 15$	*[1]*
$3 \times 6 = 18$	*[1]*
2. $\frac{1}{16} \times 2\frac{1}{2}$	*[1]*
3. £6.54	*[1]*
4. (a) £26.28	*[1]*
(b) 470 units	*[1]*
5. $\frac{1}{2} n + 3$	*[2]*
6. (a) 7	*[1]*
(b) 2	*[1]*

Appendix 2 – Answers to attainment tests

  (c) 11   [1]
  (d) –6   [1]
  (e) 30   [1]
7. $B = \frac{A}{20}$ [2]
8. (3, 2.5)   [2]
9. 7 km²   [1]
10. $\frac{20}{25}$ [1]

          [Total = 19; Pass Mark = 14]

Test 4/5

1. (a) AD   [1]
 (b) AB   [1]
2. (a) acute   [1]
 (b) reflex   [1]
 (c) obtuse   [1]
3. 140   [1]
4. 90° (a right angle)   [1]
5. Each angle should be 60°, each side 6 cm in your diagram   [3]
6.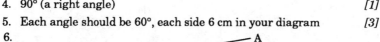
          [2]
7.
          [2]
8. 90°   [2]
9. Square-based pyramid   [1]
10. 52 cm²   [3]
11. (a) 2   [1]
  (b) 4   [1]
  (c) 6   [1]
  (d) Infinite   [1]
12.

       N
            × Appleton
       080°
     Barbridge   4 cm   [3]

13. 5 m   [1]

          [Total = 28; Pass Mark = 20]

Appendix 2 – Answers to attainment tests

Test 4/6

1. (a) Tally || ||||| ||||| ||||| ||| [2]
   Frequency   2   5   5   5   3   [1]
   (b) 5.1 [2]
   (c) 4 [1]
   (d)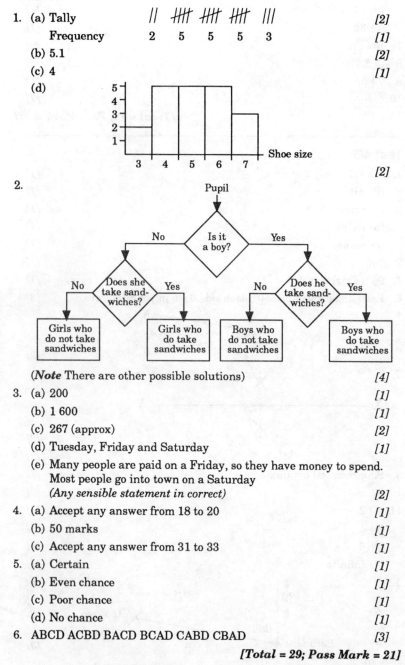
   [2]

2. (*Note* There are other possible solutions) [4]

3. (a) 200 [1]
   (b) 1 600 [1]
   (c) 267 (approx) [2]
   (d) Tuesday, Friday and Saturday [1]
   (e) Many people are paid on a Friday, so they have money to spend. Most people go into town on a Saturday
   (Any sensible statement in correct) [2]

4. (a) Accept any answer from 18 to 20 [1]
   (b) 50 marks [1]
   (c) Accept any answer from 31 to 33 [1]

5. (a) Certain [1]
   (b) Even chance [1]
   (c) Poor chance [1]
   (d) No chance [1]

6. ABCD ACBD BACD BCAD CABD CBAD [3]

[Total = 29; Pass Mark = 21]

Appendix 2 – Answers to attainment tests

Test 5/1

1. 7^4 [1]
2. 216 [1]
3. 3^5 by 118 [2]
4. 16 (*or* 4^2) [1]
5. 12 cm [3]
6. (a) 1 : 2 000 (allow 1 cm = 20 m) [1]
   (b) 60 m [1]
7. 20 cm (*or* 0.2 m) [1]
8. £2.50 [1]
9. (a) 25% [1]
   (b) 36 [1]
10. 3.5 m [1]
11. £25.20 [1]
12. $\frac{3}{10}$ [1]
13. 210 m [1]
14. 40% [1]

[Total = 19; Pass Mark = 14]

Test 5/2

1. (a) –5°C [1]
   (b) 2°C [1]
2. (d) The temperature has fallen by 8°C [1]
3. 4 500 m [1]
4. (a) 740 [1]
   (b) 9 000 [1]
   (c) 3.8 [1]
   (d) 0.4 [1]
   (e) 36.85 [1]
   (f) 7.9 [1]
   (g) 7.91 [1]
5. 5.196 (Clear working would be needed for full marks) [3]
6. 23p [1]
7. (c) 3.8287 [1]
8. 6.6 m [1]
9. 55 kg [1]
10. 6.68 m (*or* 668 cm) [1]
11. 1 600 cm^3 [1]
12. 2 kg 250 g [1]

[Total = 21; Pass Mark = 15]

Appendix 2 – Answers to attainment tests

Test 5/3

1. 69 inches [1]
2. 2 lb [1]
3. 64 km [1]
4. 0.5 litres [1]
5. 16 stones 1 pound [1]
6. Your diagram should have these measurements:

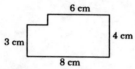

[1]

7. (a) 1 cm represents 1 km [1]
   (b) 3.5 km [1]
8. (a) 128 m [1]
   (b) Accept any answer from 6.8 cm to 6.9 cm [1]
   (c) Accept any answer from 136 m to 138 m [1]
   (d) 6 400 m² [1]

[Total = 12; Pass Mark = 9]

Test 5/4

1. 11, 13, 17, 19 [1]
2. 29 [1]
3. (a) 21 [1]
   (b) 8 [1]
4. (a) 49 [1]
   (b) 34 [1]
   (c) 216 [1]
   (d) 29 [1]
5. (a) 4 [1]
   (b) 61 [1]
   (c) 211 [1]
6. (a)

   ```
         •
        • •
       • • •
      • • • •
     • • • • •
   ```
   [1]

   (b)
   ```
   • • • •
   • • • •
   • • • •
   • • • •
   ```
   [1]

   (c)
   ```
   • • • • • •
   • • • • • •
   • • • • • •
   • • • • • •
   • • • • • •
   • • • • • •
   ```
   [1]

Appendix 2 – Answers to attainment tests

7. 27 [1]
8. 529 (*or* 23²) [2]

[Total = 17; Pass Mark = 12]

Test 5/5

1. (a) 2 [1]
    (b) 9 [1]
    (c) 2 [1]
    (d) 36 [1]
    (e) 0 [1]
    (f) 16 [1]
2. 40 m² [1]
3. (a) (4,–2) [1]
    (b) Square [2]
4. (a) 17x p + 14y p *or* (17x + 14y) pence [1]
    (b) 0.17x + 0.14y [1]
5. (a) Xr pence + Ys pence *or* (Xr + Ys) pence [1]
    (b) 90p [1]

[Total = 14; Pass Mark = 10]

Test 5/6

1. A and B [1]
2. A and C [1]
3. 36 [1]
4. a = 130 [1]
    b = 50 [1]
    c = 130 [1]
5. 40 [1]
6. 110 [1]
7. 70 [2]
8. AB and CD are parallel [1]
9. (a) 100 [1]
    (b) 100 [1]
    (c) 80 [1]
10. (a) 20 [1]
    (b) 80 [1]
    (c) 100 [1]
    (d) 40 [1]
    (e) 40 [1]
11. (a) 4 [1]

Appendix 2 – Answers to attainment tests

    (b) 2 *[1]*
    (c) 0 *[1]*
    (d) Infinite *[1]*
12. S to A to E to B to C to D to S (*or* the opposite way) *[2]*
    Distance 20 km *[2]*

[Total = 27; Pass Mark = 19]

Test 5/7

1. 26° *[1]*
2. 151° *[1]*
3. 219° *[1]*
4. 328° *[1]*
5. x = 31° (*allow an error of 2°*) *[1]*
   y = 102° (*allow an error of 2°*) *[1]*
   z = 47° (*allow an error of 2°*) *[1]*
6. (b) 80° (*marks include 2 marks for part (a)*) *[3]*
   (c) Accept any answer from 12.1 cm to 12.3 cm *[1]*
7. 32 cm *[1]*
8. 5 cm *[1]*
9. 20 cm *[1]*

[Total = 14; Pass Mark = 10]

Test 5/8

1. (a) Tally ||| ||||| | ||||| |||| | | *[2]*
    Frequency   3   6   5   4   1   1 *[2]*

   (b) [histogram: bars at heights 3, 6, 5, 4, 1, 1 over intervals 1.4–1.5, 1.5–1.6, 1.6–1.7, 1.7–1.8, 1.8–1.9, 1.9–2.0; x-axis labelled "Height in metres"] *[2]*

2. (a) 210 *[1]*
   (b) 160 *[1]*
   (c) 140 *[1]*

3. (a)

BBC 1	42	126°
BBC 2	23	69°
ITV	44	132°
Channel 4	11	33°
Total	120	360°

[1]
[1]
[1]
[1]
[1]

Appendix 2 – Answers to attainment tests

(b) A pie chart should be drawn.

A protractor should then be used to measure the angles

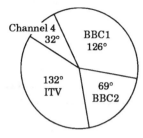

[3]

4. (b) (i) £1.50 [3]

(The marks include two marks for your graph in part (a))

(ii) 7.5 German marks [1]

5.
Weight	Tally	Frequency	
35 kg and less than 40 kg	ℋℋ ℋℋ	10	[2]
40 kg and less than 45 kg	ℋℋ //	7	[2]
45 kg and less than 50 kg	///	3	[2]

6. (a) 30 [1]
   (b) 28 [2]
7. (a) See solution to Project 12, Part a (page 136) [3]
   (b) $\frac{1}{9}$ (or $\frac{4}{36}$) [1]
   (c) 7 [1]
   (d) $\frac{1}{6}$ (or $\frac{6}{36}$) [1]
8. $\frac{1}{11}$ [1]

[Total = 37; Pass Mark = 26]

Test 6/1

1. (b) 2 tenths [1]
2. (d) 3 thousandths [1]
3. (c) 23 hundredths [1]
4. 0.0783, 0.098, 0.109, 0.1103 [2]
5. (a) 6 [1]
   (b) 12 [1]
6. $\frac{7}{10}$ [1]
7. 0.375 [1]
8. 20% [1]
9. 38.5% [1]
10. $\frac{27}{40}$ [1]
11. 43.75% [1]
12. £15 [1]

Appendix 2 – Answers to attainment tests

13. 13	*[1]*
14. 62.5	*[1]*
15. 20%	*[1]*
16. £18	*[1]*
17. £17 712	*[1]*
18. £60	*[1]*

[Total = 20; Pass Mark = 14]

Test 6/2

1. (a) 1 : 5	*[1]*
(b) 5 : 4	*[1]*
(c) 2 : 3 : 4	*[1]*
(d) 2 : 1	*[1]*
2. 36	*[1]*
3. £10 000	*[2]*
4. (a) 5 oz of flour, $\frac{3}{4}$ pint of milk, 1 egg	*[3]*
(b) 25 oz of flour, $3\frac{3}{4}$ pints of milk, 5 eggs	*[3]*
5. 54%	*[1]*
6. 25%	*[1]*
7. 17%	*[1]*
8. 30%	*[1]*
9. (b) 7	*[1]*
10. (d) 1 800	*[1]*
11. (c) 176 cm	*[1]*
12. £16	*[1]*

[Total = 21; Pass Mark = 15]

Test 6/3

1. 351, 520	*[1]*
2. 2 919, 8 751	*[1]*
3. (a) 7	*[1]*
(b) 9.5	*[1]*
(c) −5.5	*[1]*
(d) 4	*[1]*
(e) 10	*[1]*
4. (a) 2.646 (Working should be shown)	*[3]*
(b) 2.289 (Working should be shown)	*[3]*
5. The graph should be a straight line from the point (−3,−1) to (3,5)	*[4]*
6. Values of y are 3, 2, 1, 0, −1, −2, −3	*[2]*

Appendix 2 – Answers to attainment tests

7.
x	−4	−3	−2	−1	0	1	2	3	
$0.5x^2$	8	4.5	2	$\frac{1}{2}$	0	$\frac{1}{2}$	2	4.5	[3]
+2	+2	+2	+2	+2	+2	+2	+2	+2	[2]
y	10	6.5	4	2.5	2	2.5	4	6.5	[2]

The graph should go through the following points
(−4,10), (−3,6.5), (−2,4), (−1,2.5), (0,2),
(1,2.5), (2.4) and (3,6.5) [4]

[Total = 30; Pass Mark = 21]

Test 6/4

1. 2 squares and 3 triangles [1]
2. The net should have three rectangles, each 3 cm by 2 cm, and two equilateral triangles, each side 2 cm, each angle 60° [3]
3. Triangular based pyramid [1]
4. (a) Rhombus [1]
   (b) Trapezium [1]
   (c) Parallelogram [1]
5. (a) If part (b) is correct assume this is correct [1]
   (b) Accept any answer from 32 km to 34 km [2]
   (c) Accept any answer from 270° to 272° [3]
   (d) Accept any answer from 090° to 092° [1]
6. (a) Accept any answer from 301° to 303° [1]
   (b) 160 km [1]
   (c) Assume this is correct if part (d) is correct [1]
   (d) There are two possible answers. Accept either for both marks
   Any answer from 060° to 062° *or* 179° to 181° [2]
7. To check the answer place a mirror along the line XY.
   The mirror image and the drawing should be the same [1]
8. A = 1 [1]
   C = 1 [1]
   H = 2 [1]
   Z = 0 [1]
9. Scale factor of 3 [1]
10. Co-ordinates are (6,3), (10,3), (6,5) and (10,5) [4]
11. On the enlarged triangle   AB' should be 4 cm [1]
                                AC' should be 6 cm [1]
                                The area is 12 cm^2 [2]
12. 201 m^2 (approx) [2]

[Total = 36; Pass Mark = 25]

Appendix 2 – Answers to attainment tests

Test 6/5

1. (a) Positive correlation *[1]*
   (b) No correlation *[1]*
   (c) Negative correlation *[1]*
2. Your scatter diagram should show positive correlation, the crosses should extend from bottom left to top right *[3]*
3.

Name \ Hours	1	2	3	4	5	6
Adam					✓	
Brendan		✓				
Carolyn				✓		
Deborah			✓			
Elaine	✓					
Francis						✓
Gordon				✓		
Hannah			✓			

[1]

4. (a) £0.39 *[1]*
   (b) £70 *[1]*
   (c) (i) £4.42 *[1]*
       (ii) £119.68 *[1]*
5.

B	0	0	1	1
C	1	1	0	1
D	1	0	1	0

[1]
[1]
[1]

6. (a)

   (b) A to B to D (This is quicker than A to C to B to D) *[1]*
   [1]

7. The outcomes for your tree diagram should be

   HHHH, HHHT, HHTH, HHTT,
   HTHH, HTHT, HTTH, HTTT,
   THHH, THHT, THTH, THTT,
   TTHH, TTHT, TTTH, TTTT *[2]*

   (a) $\frac{3}{8}$ (or $\frac{6}{16}$) *[3]*
   (b) $\frac{11}{16}$ *[3]*

8. (a) (i) $\frac{2}{5}$ (or $\frac{4}{10}$) *[1]*
       (ii) $\frac{1}{2}$ (or $\frac{5}{10}$) *[1]*
       (iii) $\frac{1}{10}$ *[1]*
       (iv) $\frac{3}{5}$ (or $\frac{6}{10}$) *[1]*

[Total = 30; Pass Mark = 21]

Appendix 2 – Answers to attainment tests

Test 7/1

1. (a) $3 \times 3 \times 5 \times 5$ (or $3^2.5^2$) [1]
   (b) $2 \times 2 \times 3 \times 5 \times 5$ (or $2^2.3.5^2$) [1]
   (c) $2 \times 2 \times 2 \times 3 \times 3 \times 7$ (or $2^3.3^2.7$) [1]
2. 36 [2]
3. 1 120 [2]
4. 800 apples [1]
5. (a) (i) £12.58 [1]
       (ii) £47.02 [1]
   (b) (i) 39.26 D.M. [1]
       (ii) 85.59 D.M. [1]
6. (a) 2.68 kg (correct to 2 decimal places) [1]
   (b) 17.424 lb [1]
7. (a) 0.286 [1]
   (b) 8.6776 [1]
8. A = 3 [1]
   B = 4 [1]
   C = 4 [1]
9. (a) 351 km/h [2]
   (b) 97.5 m/s [3]

[Total = 24; Pass Mark = 17]

Test 7/2

1. (a) 75.79 km/h [2]
   (b) 80 km/h [1]
2. 28 minutes [2]
3. 456 kg (or 456 000 g) [3]
4. 0.85 g/cm³ [3]
5. 61 km/h [2]
6. 68.57 km/h (correct to 2 decimal places) [2]
7. (a) 36 mm [1]
8. (a) kilometres [1]

[Total = 17; Pass Mark = 12]

Test 7/3

1. 0, 7, 26, 63 [2]
2. (a) $2n + 1$ [2]
   (b) $n^2 + 6$ [2]
   (c) $\dfrac{n + 2}{n + 3}$ [2]

Appendix 2 – Answers to attainment tests

3. $\frac{1}{5}$ (*or* 0.2) [1]
4. (a) $\frac{1}{7}$ [1]
   (b) 8 [1]
   (c) $\frac{4}{3}$ (*or* $1\frac{1}{3}$) [1]
5. 11 [1]
   −12 [1]
6. (a) a^7 [1]
   (b) $8a^8$ [1]
   (c) $5y^2$ [1]
   (d) $27y^6$ [1]
7. −3, −2, −1, 0, 1 [1]
8. $x = 2.37$ (Working must be shown)
   Another possible solution is −3.37 [3]

[Total = 22; Pass Mark = 16]

Test 7/4

1. $a = 1, y = 3$ [4]
2. $a = 3, c = 2$ [4]
3. (a) 13.5 litres [1]
   (b) About 3.1 gallons [1]
   (c) 20 miles [2]
4. $x = 3, y = 1$ [4]
5. Maximum 22.5625 m^2 [1]
   Minimum 21.6225 m^2 [1]

[Total = 18; Pass Mark = 13]

Test 7/5

1. 7.21 m (correct to 2 decimal places) [1]
2. 8.94 cm (correct to 2 decimal places) [1]
3. 76.32 m (correct to 2 decimal places) [1]
4. The locus is a circle radius 6 cm [1]
5. Dotted line indicates the locus

   A ---- 20° / 20° — B (5 cm), C (5 cm) [1]

6. This diagram is not drawn to scale

   Semi-circle radius 3 cm; Semi-circle radius 3 cm; Quarter circle radius 3 cm; 3 cm; 4 cm; 7 cm [1]

Appendix 2 – Answers to attainment tests

7. The co-ordinates of S' on your diagram are
   (3,1), (4,1), (4,2) and (3,2) [1]
8. (a) 24 cubic units (accept 24) [1]
   (b) 52 square units (accept 52) [1]
   (c) (1.5,1,2) [1]
9. 42.4 cm³ [2]

[Total = 12; Pass Mark = 8]

Test 7/6

1. (a)

Interval	Tally	Frequency	Mid-point
1.40 ≤ H < 1.45	//	2	1.425
1.45 ≤ H < 1.50	//	2	1.475
1.50 ≤ H < 1.55	////	5	1.525
1.55 ≤ H < 1.60	///	3	1.575
1.60 ≤ H < 1.65	////	4	1.625
1.65 ≤ H < 1.70	////	4	1.675

[1]

   (b) 1.5675 m [2]
2. (a) 7 [1]
   (b) 7 [1]
   (c) 6 [1]
   (d) 5.636 (correct to 3 decimal places) [1]
3. (a) and (b) A line graph showing the frequency distribution
       for the men and women [2]
   (c) The graph shows that the men were generally older
       than the women [1]
4. If your flow diagram produces an answer of 324 assume that
   it is correct. The flow diagram must contain a loop [3]
5. The line of 'best fit' should go from top left to bottom right. [1]
   The diagram shows negative correlation [1]
6. (a) $\frac{1}{6}$ (or $\frac{3}{18}$) [1]
   (b) $\frac{7}{18}$ [1]
   (c) $\frac{2}{3}$ (or $\frac{12}{18}$) [1]
   (d) 1 (or $\frac{18}{18}$) [1]
   (e) 0 [1]

[Total = 25; Pass Mark = 18]

Appendix 3 – One-hour tests

Level 5: Test 1

Solutions on page 173 *Time: One hour*

1. Express $3 \times 3 \times 3 \times 4 \times 4$ using index notation.
2. This is a scale drawing of a room. The distance from A to B is 2 m.

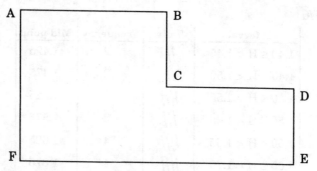

   (a) What is the scale of the drawing?
   (b) What is the distance from C to D in the room?
   (c) What is the perimeter of the room?
   (d) What is the area of the room?
3. A dress which normally sells for £18 is reduced by 10% in a sale. What is the sale price?
4. A jug contains 1.5 litres of water. This is poured into two containers so that one holds twice as much as the other. How much water does each container hold?
5. Jayne bought a radio costing £76. Her parents agreed to pay three-quarters of the money and Jayne paid the rest. How much did Jayne pay?
6. Add these numbers and write your answer correct to two significant figures.

   $$78.36 + 285.29$$
7. A lift is on floor 8. It goes down 3 floors and up 7 floors. It is now at the top floor of the building. How many floors does the building have?
8. What is the next prime number after 13?
9. Complete the next two number in these sequences.

   (a) 4, 7, 10, 13, 16, ...
   (b) $1, \frac{1}{2}, \frac{1}{4}, \frac{1}{8}, \frac{1}{16},$ ......
   (c) 1, 4, 9, 16, 25, 36, ...

Appendix 3 – One hour tests

10. What is the special mathematical name given to these numbers?

    1, 3, 6, 10, 15, 21, ……

11. How many cubes are needed to make this shape?

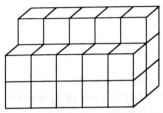

12. What is the value of 2a + 3b when a = 3 and b = 10?

13. A French man is 1.77 m tall. An English man is 5 feet 8 inches tall. Who is taller and by how many centimetres? (1 inch = 2.5 centimetres.)

14. (a) Give the co-ordinates of the point A.

    (b) The point B is 5 to the left and 4 down from A. What are the co-ordinates of the point B?

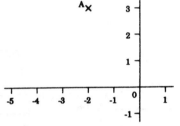

15. Mrs Elliot requires 28 litres of paint. Paint is sold in the following sized tins. 10 litre tins cost £18. 5 litre tins cost £10. 1 litre tins cost £2.75. How many tins of each size should she buy to minimise her cost?

16. Use your protractor to measure these angles.

    (a)   (b)   (c)

17. What is the value of x?

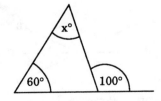

Appendix 3 – One hour tests

18. Find the value of x.

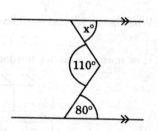

19. Copy these shapes, mark all of the axes of symmetry with dotted lines, state the number of axes of symmetry and give the mathematical name of each shape.

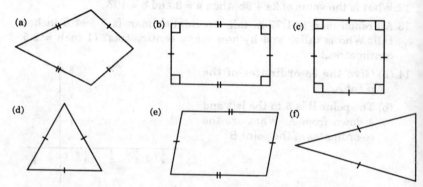

20. The heights of 25 children are shown below. Heights are in centimetres.

```
137  152  173  141  147  163  158  177  162  139
140  152  161  147  173  138  164  171  149  152
171  157  163  159  147
```

Using equal class intervals copy and complete this table.

Height in centimetres (H)	Tally	Frequency
$135 \leq H < 140$	///	3
$140 \leq H < 145$		

21. AD, BE and CF are straight lines. What is the size of angle EOD when $x = 40°$?

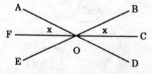

22. In a test Sarah scored 60 marks. Her percentage score was 75%. John scored 48 marks. What was his percentage score?

Appendix 3 – One hour tests

Level 5: Test 2

Solutions on page 174   *Time: One hour*

1. What is the value of 52.33?
2. A rectangular garden is 18 m by 12 m.
   (a) Draw an accurate scale drawing of the garden using a scale of 1 cm represents 3 m.
   (b) What is the diagonal length of the garden?
3. 15% VAT is added to the price of a meal costing £30.
   (a) How much VAT is added?
   (b) What is the cost of the meal including VAT?
4. A man normally earns £4.40 per hour. On Tuesday he worked his normal 8 hour shift plus 2 hours overtime at time and a half. How much did he earn on Tuesday?
5. Express 80 pence as a fraction of £5.
6. Write the answer to this question correct to one decimal place. 78.239 – 17.446.
7. Find the square root of 20 correct to two decimal places without pressing the square root key ($\sqrt{\ }$) on your calculator. Show all of your working.
8. What is the cube root of 64?
9. Find the value of $\sqrt[3]{216} - \sqrt[3]{27}$.
10. These are square numbers. 16, 25, 36, 49. Write down the corresponding cube numbers.
11. Fill in the missing numbers in these sums.

    (a)
    ```
        6 ? 3
      + ? 4 ?
      -------
      1 0 6 0
    ```
    (b)
    ```
        3 ? 4
      ×     2
      -------
        ? 4 ?
    ```

12. What is the perimeter of this rectangle when x = 5 m and y = 3 m?

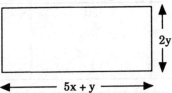

13. A tractor (T) length x metres is pulling two wagons (W), each y metres long.

    (a) Express the total length of the tractor and wagons mathematically.

The tractor is one and a half times the length of each wagon. The total length of the tractor and wagons is 13.5 metres.

(b) What is the length of the tractor correct to the nearest centimetre?

14. A car can travel 8 miles on one litre of petrol. How many gallons would be needed to travel 360 miles? (1 gallon = 4.5 litres.)

15. Bacon-flavoured crisps are sold in 80 gram bags for 14p each.

    (a) What is the cost per kilogram?

    Beef-flavoured crisps are sold in 100 gram bags for 15p each.

    (b) Which flavour is cheaper and by how much per kilogram?

16. Construct the triangle, AB = 8 cm, BC = 6 cm, AC = 5 cm. Measure and write down the size of angle ABC.

17. What is the value of x?

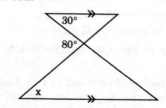

18. What is the value of x?

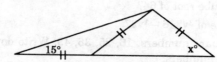

19. Ninety people were asked what was their favourite season. Their answers are shown in the table below. Copy and complete the table below and hence draw a pie chart to show the information.

Favourite Season	Number of people	Angle at the centre of the pie chart
Spring	30	
Summer	38	
Autumn		
Winter	5	

20. Pupils arrive at school early, on time or late. On January 27th a group of pupils were observed. From these observations the following probabilities were found. The probability of a pupil arriving early was $\frac{2}{5}$. The probability of a pupil arriving on time was $\frac{1}{2}$.

    (a) What was the probability of a pupil arriving late?

    (b) In the statistics used to determine these probabilities 80 pupils arrived early.

    (i) How many pupils were observed altogether?

Appendix 3 – One hour tests

  (ii) How many children arrived late?

  (iii) What percentage of the pupils arrived early?

21. This pie graph represents the number of boys and girls in a school.

    (a) What fraction of the pupils are girls?

    (b) What percentage of the pupils are boys?

    (c) 1 000 pupils attend the school.

      (i) How many girls attend the school?

      (ii) How many boys attend the school?

    (d) 8% of the pupils are sixth formers. How many sixth formers attend the school?

22. The mean average of three numbers is 8. Two of the numbers are 6 and 7. What is the third number?

Appendix 3 – One hour tests

Level 5: Test 3

Solutions on page 175 *Time: One hour*

1. What is the difference between 7^3 and 5^3?
2. A plan of a room is drawn to a scale of 1 : 300. The length of the room on the plan is 4cm. What is the length of the actual room?
3. A box contained 160 eggs. 5% were bad. How many good eggs were in the box?
4. Express a quarter of a kilogram as a fraction of two kilograms. Give your answer in its lowest terms.
5. The temperature at 1200 on Christmas day was 5°C. By 1700 it had fallen by 7°. What was the temperature at 1700?
6. The local paper stated that 18 000 people attended a football match. If the figure is correct to the nearest thousand, what is the lowest number of people who could have attended?
7. A man received a garage bill for £23.72 + 15% VAT. Calculate the VAT correct to the nearest penny.
8. List all of the factors of 12.
9. Work through this flow chart. Start with N = 1000.

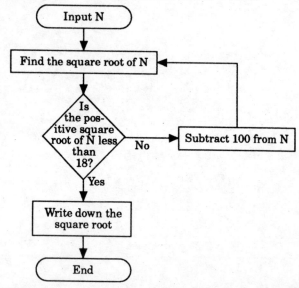

10. What is the lowest cube number which is greater than 500?
11. $y = 3 + 2a$. What is the value of a when $y = 15$?
12. A water tank is 60 cm long, 40 cm wide and 50 cm high. What is the capacity of the tank in (a) cubic centimetres (b) litres?
13. Petrol is sold at 40p a litre. My car travels 32 miles on one gallon of petrol. What is the cost of travelling 160 miles? (1 gallon = 4.5 litres)

Appendix 3 – One hour tests

14. What is the smallest number, which is greater than 14, which can be divided exactly by 3 and 4?

15. The area of a trapezium is calculated by the formula Area = $\frac{1}{2}$ (a + b) × h.
What is the area of this trapezium?

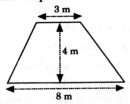

16. Subtract 2 320 cm from 0.78 km and give your answer in metres.

17.

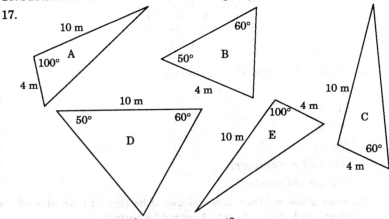

(a) Which two triangles are congruent?
(b) Which two triangles are similar (but not congruent)?

18. These are the recorded speeds (in miles per hour) of 20 cars on a motorway.
    52.7 69.3 71.3 76.7 61.3 48.4 72.4 67.3 68.4 58.3
    64.2 56.3 49.8 50.2 47.6 72.8 63.5 62.9 60.0 71.3

(a) Copy and complete the table below.

Speed (S) in miles per hour	Tally	Frequency
40 ≤ S < 50		
50 ≤ S < 60		
60 ≤ S < 70		
70 ≤ S < 80		

(b) Draw a frequency graph to show this information.
(c) What fraction of the cars were travelling at less than 50 miles per hour?
(d) The speed limit on the motorway is 70 miles per hour.

Appendix 3 – One hour tests

   (i) What percentage of the cars were breaking the speed limit?

   (ii) By how many miles per hour was the fastest car breaking the speed limit?

19. (a) Draw a conversion graph to convert gallons to litres (2 gallons = 9 litres).

   (b) Use your graph to answer the following questions.

   (i) Convert 3.5 gallons to litres.

   (ii) How many gallons are equivalent to 13.5 litres?

20. John decided to alter the numbers on his die. The net of the die after the numbers had been changed is shown here.

		6	
3	2	3	1
		2	

   (a) What is the probability of shaking:

   (i) 1;

   (ii) 2;

   (iii) 3;

   (iv) 5;

   (v) an even number;

   (vi) an odd number?

   (b) John made another identical die. If he throws both dice which numbers between 2 and 14 cannot be scored?

21. What is the value of x?

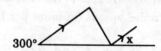

Appendix 3 – One hour tests

Level 6: Test 1

Solutions on page 176   *Time: One hour*

1. What is the value of the 8 in the number 3.587?
2. Fill in the missing numbers
$$\frac{4}{5} = \frac{?}{10} = \frac{32}{?}$$
3. Which is worth most and by how much? 80% of £16 or $\frac{2}{3}$ of £22.50.
4. The normal price of a television set is £240. The normal price of a radio is £54. These items can be bought at a 20% discount in the sale, or the customer can choose the special offer and pay the full price for the television and receive the radio free of charge. Mrs Garner decides to purchase the television and radio. Which method will be cheaper and by how much?
5. A washing machine normally costs £280. In a sale there is a reduction of 18%. What is the sale price?
6. Some money was divided between Anna and Brenda in the ratio 3 : 5. Brenda received £150 more than Anna. How much did Anna receive.
7. A zoo had 320 monkeys. 5% were apes. How many apes were there?
8. 8 men can build a house in 25 days. How long would 10 men take?
9. Solve these equations.
   (a) 3 + 2y = 17
   (b) y – 3 = 2
   (c) $\frac{8y}{3}$ = 36.
10. Solve this equation using a trial and improvement method. Produce an answer which is correct to 3 decimal places. Show all working.
$$x^2 = 11.$$
11. Draw the line y = 2x – 1 for the values –2 ≤ x ≤ 4.
12. A train leaves Stoke Station at 09.38 and arrives in London at 12.15. The distance from Stoke to London is 140 miles.
    (a) How long did the journey take?
    (b) What was the speed of the train correct to the nearest 10 miles per hour?
13. This is a block of metal.
    The weight of the block is 8.304 kg.
    What is the density of the metal?
14. How many axes of symmetry does each of these shapes have? Draw the shapes and axes.
    (a) Square.   (b) Trapezium.   (c) Rhombus.
15. Draw the net of a cube, side 3 cm.

Appendix 3 – One hour tests

16. What is the mathematical name of this shape?

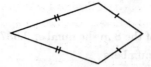

17. A plane starts from A. It flies on a bearing of 136° for 45 minutes at a speed of 400 km/h to B. It then changes course to 080° and flies for 15 minutes at a speed of 300 km/h to C.
    (a) Using a scale of 1 cm represents 50 km draw a scale diagram to show the flight of the plane.
    (b) What is the distance from A to C?
    (c) What bearing must the plane fly to return to A?
    (d) The plane flies directly from C to A at an average speed of 400 km/h. How long does the flight take? Give your answer in minutes, correct to the nearest minute.

18. This shape is symmetrical about the y axis. Draw and complete the shape.

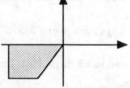

19. Draw an accurate enlargement of this rectangle, scale factor 3.

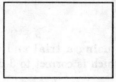

20. The probability of a wet day is 0.35. The probability of a dry day is 0.65.
    (a) Copy and complete this tree diagram.

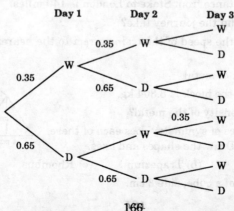

Appendix 3 – One hour tests

(b) What is the probability of two consecutive wet days?

(c) What is the probability of three consecutive days without rain?

(d) What is the probability that the second day is dry?

21. One lap of a running track measures 400 m.
    (a) How many laps is the 10 000 m race?
    (b) The fastest runner completes the race in exactly 32 minutes. What was his average speed in:
        (i) metres per minute?
        (ii) metres per second?
        (iii) kilometres per hour?

22. Mrs Miller earns £12 000 pa. 6% National Insurance is deducted.
    (a) How much National Insurance does she pay?
    She pays tax of 25% on the remainder of her salary.
    (b) How much tax does she pay?
    (c) How much money does she actually receive each year?
    (d) What is her take-home pay each week, correct to the nearest penny? (52 weeks in one year.)

23. AD, BE and CF are straight lines. Write down the size of Angle EOD in terms of x.

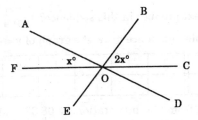

24. All triangles in these diagrams are equilateral, the small triangles are all the same size.

Shape A   **Shape B**

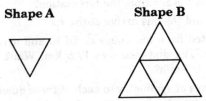

(a) Write down the perimeter of shape A to the perimeter of Shape B as a ratio.

(b) Write down the area of Shape A to the area of Shape B as a ratio.

Appendix 3 – One hour tests

Level 6: Test 2

Solutions on page 178 *Time: One hour*

1. 0.07 is equal to how many thousandths?
2. Which is the greater and by how much, 0.675 or $\frac{11}{16}$?
3. Reduce £75 by $\frac{2}{3}$.
4. A meal cost £7.60. 15% VAT was added to this. What was the total cost of the meal?
5. £5 000 was divided between three brothers Alan, Brian and Clive in the ratio 5 : 3 : 2. How much did each brother receive?
6. 28 pupils out of a class of 32 were present on Tuesday. What percentage is this?
7. A dog kennel bred 375 dogs. 25 of these were terriers. What percentage is this? Give your answer correct to two decimal places.
8. 20 bars of chocolate cost £2.40. What is the cost of 15 bars of chocolate?
9. Solve these equations.
   (a) $\frac{x}{3} = 8$
   (b) $5x - 2 = 3x + 15$
   (c) $2(x + 3) = 10$
10. What is the next number in this sequence? $1, \frac{1}{3}, \frac{1}{9}, \frac{1}{27}, \frac{1}{81}, \ldots$
11. Fill in this table and hence draw the graph of $y = -\frac{x}{2}$.

x	−2	−1	0	1	2	3
y						

12. A bus started from the bus station at 08.37 and travelled to the zoo. The bus maintained an average speed of 78 km/h and the journey took 1 hour 40 minutes.
    (a) How far was the zoo from the bus station?
    (b) What time did the bus arrive at the zoo?
13. Mr Allen started from his house at 10.40 and arrived at work 12 minutes later. The distance was 12.5 km. What was Mr Allen's average speed in km/h?
14. Draw in the axes of symmetry in each of these quadrilaterals.

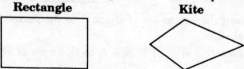

15. All four sides of this shape are equal lengths. What is the name of this shape?

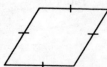

Appendix 3 – One hour tests

16. This is the net of a cuboid box. The volume of the box is 6 m³. What is
    (a) the surface area of the base?
    (b) the total surface area of the box?

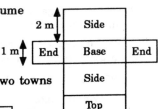

17. This diagram shows the positions of two towns A and B. (Scale 1 cm = 6 km)

    (a) How far is A from B?
    (b) What is the bearing of B from A?
    (c) What is the bearing of A from B?

18.   **B C E F I N**

    Which of these letters have:
    (a) no lines of symmetry?
    (b) one line of symmetry?
    (c) two lines of symmetry?

19. This table shows the monthly repayments on a mortgage.

Amount borrowed £'s	Interest Rate of 11% £'s	Interest Rate of 11.5% £'s
100	0.92	0.96
200	1.84	1.92
300	2.75	2.88
400	3.67	3.84
500	4.59	4.80
1 000	9.17	9.59
2 000	18.34	19.17
3 000	27.50	28.75
4 000	36.67	38.34
5 000	45.84	47.92
10 000	91.67	95.84

(a) What is the cost of borrowing £3 400 at 11%?
(b) Mr Jones buys a house for £19 200. He borrows 75% of the money from the building society. What is the cost of borrowing this money at 11%
   (i) per month?
   (ii) per year?

Appendix 3 – One hour tests

(c) The mortgage rate increased to 11.5%. How much more money did Mr Jones have to pay each month?

20. A bag contains 3 red balls, 2 blue balls and 1 black ball.
    (a) What is the probability of selecting a red ball?
    (b) What is the probability of selecting a blue ball?
    (c) How many more black balls would have to be placed in the bag to make the probability of choosing a black ball $\frac{2}{3}$?

21. The price of a car is £7 500. On to this price 17.5% VAT is added. What is the total price of the car?

22. This pie chart represents the number of male and female workers in a factory. The total number of workers is 180.

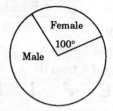

(a) What fraction of the workers is female?
(b) How many female workers are there?

Appendix 3 – One hour tests

Level 6: Test 3

Solutions on page 179   *Time: One hour*

1. Write these numbers in descending order. 0.3721, $\frac{3}{8}$, 0.42, $\frac{2}{5}$.
2. Write 85% as a fraction in its lowest terms.
3. In a sale, a scarf which normally cost £8 was reduced by 20%. Mrs Clark bought the scarf and paid with a £10 note. How much change did she receive?
4. Simplify these ratios.
   (a) 5 : 15; (b) 25 : 20; (c) $\frac{1}{2} : \frac{1}{8}$.
5. The Longshaw Mine in California produced 8 500 000 tonnes of coal last year. This was 8% of the total coal production of California. How much coal was produced in the whole of California last year?
6. What is 7 597 ÷ 382 correct to the nearest whole number?
7. Find the sum of 25% of £30 and 10% of £60.
8. Solve the equation 7(3a − 5) = 4a (correct to 2 decimal places).
9. What is the next number in this sequence? 3, 7, 15, 27, 43, 63, ...
10. Draw the graph of $y = 2x^2$ for the values $-3 \leq x \leq 3$.
11. A block of wood has a density of 0.81 g/cm3. The total weight of the block is 0.1458 kg. The area of the base is 45 cm2. What is the height of the block of wood?
12. Which metric units of length would you choose to measure the lengths of the following: (a) a fly; (b) a room; (c) a road; (d) a book? (Choose from mm, cm, m, km.)
13. This shape is a kite.
    What is the size of angle:
    (a) ADC; (b) ADB; (c) ACD?

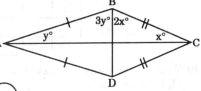

14. This is the net of a shape.

    What is the mathematical name of the shape?
15. What is the special name given to this shape?

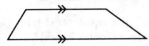

16. Triangle XYZ is an enlargement of triangle ABC. The ratio of the area of triangle ABC to triangle XYZ is 1 : 16. What is the length of XY?

Appendix 3 – One hour tests

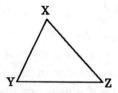

17. Radar station B is 160 km due east of radar station A. A plane is spotted on radar. The bearing is 065° from radar station A and 290° from radar station B.

    (a) Using a scale of 1 cm represents 20 km, draw a scale diagram to show the position of both radar stations and the plane.

    (b) How far is the plane from radar station A?

18. This network shows the bus routes between four towns.

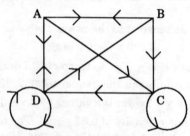

Copy and complete this table to show that information.

	A	B	C	D
A				
B				
C				
D				

19. When three coins are tossed what is the probability that they will all land on heads?

20. $y = \dfrac{1}{x} + \dfrac{1}{2x}$.

    Find the value of y when $x = \dfrac{1}{5}$.

21. This is an equilateral triangle, side 3 cm.

    (a) What is the size of each angle?

    (b) The area of the triangle is 2.598 cm². What is the perpendicular height of the triangle?

    (c) The perimeter of this equilateral triangle is 36 m. What is its perpendicular height?

    (Triangles not drawn to scale.)

Appendix 4 – Mock exam answers to one hour papers in Appendix 3

[Marks allocated to each question are given in brackets]

Level 5

Test 1

1. $3^3.4^2$ [1]
2. (a) 1 : 50 (accept 2 cm represents 1 m) [1]
   (b) 1.75 m (*or* 175 cm) [1]
   (c) 11.5 m [1]
   (d) 5.75 m^2 [2]
3. £16.20 [1]
4. 1 litre and 0.5 litres [1]
5. £19 [1]
6. 360 [1]
7. 12 floors [1]
8. 17 [1]
9. (a) 19, 22 [1]
   (b) $\frac{1}{32}, \frac{1}{64}$ [1]
   (c) 49, 64 [1]
10. Triangular numbers [1]
11. 25 [1]
12. 36 [1]
13. Frenchman is 7 cm taller [2]
14. (a) (–2,3) [1]
    (b) (–7,–1) [1]
15. Three ten-litre tins (cost £54) [2]
16. (a) Accept any answer from 142° to 144° [1]
    (b) Accept any answer from 319° to 321° [1]
    (c) Accept any answer from 46° to 48° [1]
17. 40 [1]
18. 30 [1]
19. (a) Kite   one axis   [2]
    (b) Rectangle   two axes   [2]
    (c) Square   four axes   [2]
    (d) Equilateral triangle three axes [2]
    (e) Parallelogram no axes   [2]
    (f) Isoceles triangle one axis [2]

Appendix 4 – Mock exam answers to one hour papers in Appendix 3

20.

Interval	Tally	Frequency	
135 ≤ H < 140	///	3	
140 ≤ H < 145	//	2	
145 ≤ H < 150	////	4	
150 ≤ H < 155	///	3	
155 ≤ H < 160	///	3	
160 ≤ H < 165	/////	5	
165 ≤ H < 170		0	
170 ≤ H < 175	////	4	
175 ≤ H < 180	/	1	*[4]*

21. 100° *[1]*
22. 60 *[1]*

[Total = 47; Pass Mark = 33]

Test 5/2

1. 675 *[1]*
2. (a) Your diagram should be 6 cm by 4 cm *[1]*
   (b) Accept any answer from 21 m to 22 m *[1]*
3. (a) £4.50 *[1]*
   (b) £34.50 *[1]*
4. £48.40 *[1]*
5. $\frac{4}{25}$ *[1]*
6. 60.8 *[1]*
7. 4.47 (Working must be shown) *[3]*
8. 4 *[1]*
9. 3 *[1]*
10. 64, 125, 216, 343 *[2]*
11. (a)
    ```
      6 1 3
      4 4 7
    -------
    1 0 6 0
    ```
    [2]

    (b)
    ```
    3 2 4            3 7 4
    × 2      or      × 2
    -----            -----
    6 4 8            7 4 8
    ```
    [2]
12. 68 m *[2]*
13. (a) (x + 2y) metres *[2]*
    (b) 579 cm (accept 5.79 m) *[1]*
14. 10 gallons *[1]*
15. (a) 175 p (*or* £1.75) *[1]*
    (b) Beef are cheaper by 25 p per kilogram *[2]*
16. Accept any answer from 38° to 40° for angle ABC *[3]*

Appendix 4 – Mock exam answers to one hour papers in Appendix 3

17.	50°	[1]
18.	30°	[1]
19.	Number of people for Autumn is 17	[1]
	Angles at centre are Spring = 120° Summer = 152° Autumn = 68° Winter = 20° *Your pie chart should have angles as indicated above*	[2]
20. (a)	$\frac{1}{10}$	[1]
(i)	200	[1]
(ii)	20	[1]
(iii)	40%	[1]
21. (a)	$\frac{2}{5}$	[1]
(b)	60%	[1]
(i)	400	[1]
(ii)	600	[1]
(d)	80	[1]
22.	11	[1]

[Total = 47; Pass Mark = 33]

Test 5/3

1.	218	[1]
2.	12 m (*or* 1 200 cm)	[1]
3.	152	[1]
4.	$\frac{1}{8}$	[1]
5.	−2°C	[1]
6.	17 500	[1]
7.	£3.56	[1]
8.	1, 2, 3, 4, 6, 12	[2]
9.	17.32 (correct to 2 decimal places)	[2]
10.	512 (accept 8^3)	[1]
11.	a = 6	[1]
12. (a)	120 000 cubic centimetres	[1]
(b)	120 litres	[1]
13.	£9	[2]
14.	24	[1]
15.	22 (*or* 22 m^2)	[1]
16.	756.8 m	[1]
17. (a)	A and E	[1]
(b)	B and D	[1]
18. (a)	Tally /// //// ++++ /// ////	[1]
	Frequency 3 4 8 5	[1]

Appendix 4 – Mock exam answers to one hour papers in Appendix 3

    (b)

(c)	$\frac{3}{20}$			[1]
(d)	(i)	25%		[1]
	(ii)	6.7 mph		[1]

19. (a) Assume this is correct if your answer to part (b)(i) is correct [1]
    (b) (i) Any answer from 15.5 litres to 16 litres [1]
        (ii) 3 [1]
20. (a) (i) $\frac{1}{6}$ [1]
        (ii) $\frac{1}{6}$ [1]
        (iii) $\frac{1}{3}$ (or $\frac{2}{6}$) [1]
        (iv) 0 [1]
        (v) $\frac{1}{2}$ [1]
        (vi) $\frac{1}{6}$ [1]
    (b) 10, 11, 13 [2]
21. 60° [1]

[Total = 40; Pass Mark = 28]

Test 6/1

1. 8 hundredths [1]
2. 8, 40 [1]
3. $\frac{2}{3}$ of £22.50 by £2.20 [1]
4. Discount of 20% is £4.80 cheaper [2]
5. £229.60 [1]
6. £225 [1]
7. 16 [1]
8. 20 days [1]
9. (a) y = 7 [1]
   (b) y = 5 [1]
   (c) y = 13.5 [1]
10. 3.317 (Working must be shown) [3]
11. The line should be straight and start at the point (−2,−5) and finish at the point (4,7) [2]

Appendix 4 – Mock exam answers to one hour papers in Appendix 3

12. (a) 2 hours 37 minutes [1]
    (b) 50 mph [1]
13. 173 g/cm³ [2]
14. (a) 4 [1]
    (b) 0 [1]
    (c) 2 [1]
15. The net should consist of six squares, sides 3 cm [2]
16. Kite [1]
17. (a) *This diagram is not drawn to scale*

    (diagram: N arrow at A, bearing 136°; A to B = 6 cm; N arrow at B, bearing 080°; B to C = 1.5 cm) [3]

    (b) Any answer between 340 km and 350 km
        (accept 6.8 to 7.0 cm) [1]
    (c) Any answer from 304° to 306° [1]
    (d) Accept 51 *or* 52 *or* 53 minutes [2]
18. (diagram) [1]
19. The enlargement should be 9 cm by 6 cm [1]
20. (a) Each wet day should be 0.35, each dry day should be 0.65 [1]
    (b) 0.1225 [1]
    (c) 0.27 (correct to 2 decimal places). Exact answer 0.274625 [1]
    (d) 0.65 [1]
21. (a) 25 laps [1]
    (b) (i) 312.5 metres per minute [1]
        (ii) 5.2 metres per second (correct to 1 decimal place) [1]
        (iii) 18.75 km/h [1]
22. (a) £720 [1]
    (b) £2 820 [1]
    (c) £8 460 [1]
    (d) £162.69 [1]
23. $60 - x$ *or* $180 - 3x$ *or* $360 - 6x$ [1]
24. (a) 1 : 2 (accept 3 : 6) [1]
    (b) 1 : 4 [1]

[Total = 51; Pass Mark = 36]

Appendix 4 – Mock exam answers to one hour papers in Appendix 3

Test 6/2

1. 70 thousandths [1]
2. $\frac{11}{16}$ by 0.0125 [1]
3. £25 [1]
4. £8.74 [1]
5. £2 500 : £1 500 : £1 000 [1]
6. 87.5% [1]
7. 6.67% [1]
8. £1.80 [1]
9. (a) $x = 24$ [1]
   (b) $x = 8.5$ [1]
   (c) $x = 2$ [1]
10. $\frac{1}{243}$ [1]
11. $y = 1, \frac{1}{2}, 0, -\frac{1}{2}, -1, -1.5$ [1]
    The graph should be a straight line from (–2,1) to (3,–1.5) [1]
12. (a) 130 km [1]
    (b) 10.17 [1]
13. 62.5 km/h [1]
14.

[1]

15. Rhombus [1]
16. (a) 3 cm² [1]
    (b) 22 cm² [2]
17. (a) Any answer from 21.5 km to 22 km [1]
    (b) Any answer from 105° to 107° [1]
    (c) Any answer from 285° to 287° [1]
18. (a) **F N** [1]
    (b) **B C E** [1]
    (c) **I** [1]
19. (a) £31.17 per month [1]
    (b) (i) £132.01 [2]
        (ii) £1 584.12 [1]

Appendix 4 – Mock exam answers to one hour papers in Appendix 3

	(c)	£6.01	[1]
20.	(a)	$\frac{1}{2}$ (accept $\frac{3}{2}$)	[1]
	(b)	$\frac{1}{3}$ (accept $\frac{2}{6}$)	[1]
	(c)	9	[1]
21.		£8 812.5	[1]
22.	(a)	$\frac{5}{18}$	[1]
	(b)	50	[1]

[Total = 40; Pass Mark = 28]

Test 6/3

1. $0.42, \frac{2}{5}, \frac{3}{8}, 0.3721$ [1]
2. $\frac{17}{20}$ [1]
3. £3.60 [1]
4. (a) 1 : 3 [1]
   (b) 5 : 4 [1]
   (c) 4 : 1 [1]
5. 106 250 000 [1]
6. 20 [1]
7. £13.50 [1]
8. 2.06 [1]
9. 87 [1]
10. The graph should go through these points
    (–3,18), (–2,8), (–1,2), (0,0), (1,2), (2,8), (3,18) [3]
11. 4 cm [3]
12. (a) mm [1]
    (b) m [1]
    (c) km [1]
    (d) cm [1]
13. (a) 127.5° [1]
    (b) 67.5° [1]
    (c) 30° [1]
14. Cylinder [1]
15. Trapezium [1]
16. 12 m [2]
17. (a) *The diagram is not drawn to scale*

    Triangle with vertices A, P, B; AP = 4 cm, angle A = 25°, angle B = 20°, AB = 8 cm [2]

Appendix 4 – Mock exam answers to one hour papers in Appendix 3

	(b)	80 km (approx)				*[2]*
18.		A	B	C	D	
	A	0	1	1	1	*[1]*
	B	1	0	1	0	*[1]*
	C	0	0	1	1	*[1]*
	D	1	1	0	2	*[1]*

19. $\frac{1}{8}$ *[1]*

20. 7.5 *[1]*

21. (a) 60° *[1]*
    (b) 1.732 cm *[3]*
    (c) 6.928 cm *[2]*

[Total = 44; Pass Mark = 31]

Notes

Notes

Notes

Notes